Pvt. Ltd.

YEASTS FOR BIOREMEDIATION

Nilanjana Das
School of Biosciences and Technology
Environmental Biotechnology Division
Bioremediation Laboratory
VIT University
Vellore - 632 014

2017

Studium Press (India) Pvt. Ltd.

YEASTS FOR BIOREMEDIATION

ISBN: 978-93-80012-86-5

Published by:

Studium Press (India) Pvt. Ltd.
4735/22, 2nd Floor, Prakash Deep Building
(Near Delhi Medical Association),
Ansari Road, Darya Ganj, New Delhi-110 002
Tel.: + 91-11-43240200-15 (15 lines); Fax: 91-11-43240215
E-mail: studiumpress@gmail.com

Printed at:

Printed In India

Dr. G. VISWANATHAN
Founder & Chancellor
Former Member of Parliament
Former Minister, Govt. of Tamil Nadu
President, Education Promotion Society for India,
New Delhi

Foreword-I

I am delighted to know that Dr. Nilanjana Mitra (Das) has authored the book ***'Yeasts for Bioremediation'***.

The importance of yeast in food and beverage industries has been known since 1860. Yeasts have been contributed in different fields of biotechnology and considered as model organisms. Recent studies have shown that yeasts can also play an important role towards the control of environmental pollution in addition to other biotechnological applications. This publication presents the potential role of yeasts in the field of bioremediation.

The author has contributed valuable inputs in her book which will definitely create awareness among the researchers, scientists and environmentalists towards the utility of yeasts as novel remediation agents.

Date: 10th August, 2015

(Dr. G. Viswanathan)
Chancellor
VIT University,
Vellore - 632 014

Vellore – 632 014, Tamil Nadu, India; Tel.: +91 416-224 3100, Fax: +91 416-224 3092;
E-mail: chancellor@vit.ac.in; Chennai Campus: Vandalur- Kelambakkam Road,
Chennai – 600 127, India; Tel.: +91 44-3993 1555; Fax: +91 44-3993 2555
www.vit.ac.in

www.vit.ac.in

DR. C. RAMALINGAM
Dean
School of Bio Sciences and Technology
VIT University,
Vellore - 632 014

Foreword-II

It is my pleasure to write a few lines as foreword to the book ***'Yeasts for Bioremediation'*** authored by Dr. Nilanjana Mitra (Das). Yeasts have been used as a potential resource for making bread, beer and wine from early civilization. After so many centuries and even millenia, many yeast species have been isolated and identified for their contribution in different fields of biotechnology and considered as most widely used model organisms for genetics and cell biology. The importance of yeast as a novel bioremediation agent for the control of environmental pollution will stimulate research in the field. The author has vividly described the role of many non pathogenic yeast species isolated from environment towards remediation of various pollutants such as heavy metals, synthetic dyes, petroleum hydrocarbons, pesticide and emerging contaminants like caffeine and antibiotics. The book will definitely serve as valuable guide for the researchers who are interested for innovative research in this area. I congratulate the author for her valuable and scientifically based piece of work.

Date: 10th August, 2015

(Dr. C. Ramalingam)
Dean
School of Biosciences and Technology
VIT University, Vellore - 632 014

Vellore – 632 014, Tamil Nadu, India; Fax: 0416-2243 092, 2240411;
Direct: 0416 - 2241360, 0416 - 2202575; E-mail: dean.sbst@it.ac.in; caramalingam@vit.ac.in
www.vit.ac.in

About the Author

Dr. Nilanjana Mitra (Das) (*b.* 1957); Currently serving as a Senior Professor in the School of Biosciences and Technology (SBST), Environmental Biotechnology Division, VIT University, Vellore- 632014, Tamil Nadu. She received her Bachelor of Science degree from Lady Brabourne College, Calcutta in 1977 and Master of Science degree in Agriculture from University of Calcutta in 1979. Dr. Das received DIIT (1982) and Ph.D (1990) degree from Agricultural Engineering Department, IIT Kharagpur. Later she continued her post doctoral research on Plant Biotechnology and Cryopreservation Technology in the Department of Chemical Engineering and Cryogenics Engineering Centre at IIT Kharagpur. She served as Senior Consultant at Rural Development Centre, IIT Kharagpur for several years and made a significant contribution in different areas of research. Dr. Das has published more than 100 research papers in peer reviewed International/National Journals. She was adjudged for an 'Award for Best Research Paper (2003)' from Indian Forester, Dehradun for outstanding research in the area of 'Environment and Ecology'. Dr Das participated in different scientific activities *viz.* National/International Seminars/Symposium/Workshops/Conferences etc. She served as PI and CO-PI for the research projects funded by Department of Science and Technology, Govt. of India and received prestigious Women Scientists Award in 2005. At present, Dr Das is In–Charge of Bioremediation Laboratory, SBST, VIT University, Vellore and actively involved in the area of research on Microbial Bioremediation.

Preface

Yeasts are a heterogenous group of fungi and grow in a conspicuous unicellular form. They reproduce by fission, budding or a combination of both. True yeasts reproduce sexually or by developing ascospores. At the beginning of 18th century, there was a well known yeast species named as *Saccharomyces cerevisiae* which was used as a potential resource for making bread, beer and wine. After so many centuries and even millenia, many yeast species have been isolated and identified for their contribution in different fields of biotechnology.

With all historical background and industrial importance, yeasts did not receive universal acceptance over the years as remediation agents like fungi and bacteria. It has been my endeavour to maintain the clarity of explanation which will contribute to the better understanding of yeasts and their future exploitation.

This book is mainly intended for the researchers who are pursuing research in Microbiology, Environmental Sciences and Biotechnology. The main purpose of this book is to create awareness among the inquisitive persons with a blend of modern concept of application of yeasts as bioremediation agent. The book has been drafted into ten chapters and written in a very concise and lucid manner.

The first chapter is primarily an introduction, providing a foundation for the descriptions of other chapters. Basic information on historical background of yeasts, morphotype, habitat, importance, isolation and identification techniques etc. have been provided in the second chapter. The third chapter deals with the concept of microbial bioremediation. Though many technologies are available for the clean up purposes, only few of them have been proved to be of routine application value. The choice of technology is influenced by the social, political and geographical conditions. Among all the biological techniques, bioremediation has been evolved as the most promising one because of its economical, safety and environmental features.

The fourth, fifth, sixth and seventh chapters provide the information on the role of yeasts as agents for the remediation of various pollutants *viz.* heavy metals, synthetic dyes, petroleum hydrocarbon and organochlorine pesticide from soil and water environment. The eighth and ninth chapter deal with the remediation of two emerging pollutants like caffeine and cephalosporin antibiotics which are largely being used by human almost every day and creating pollution in the environment. Chapter ten summarizes the information which has been provided in all the chapters.

Based on the information reported in this book, it appears that non pathogenic yeast species isolated from soil and aqueous environment can serve as potential candidates for remediation of various pollutants.

Bibliography at the end of the book will help enthusiastic person to search about the yeasts from various sources.

I appreciate the potentiality of my Ph.D scholars who have done lot of research involving yeasts as remediation agents for removal of pollutants. I sincerely hope that the researchers will definitely enjoy reading this book as much as I have enjoyed writing it. Suggestions for improvement from users are welcome.

Table of Contents

1

Introduction

There has been an increasing concern regarding the accumulation of various pollutants in environment which pose a real threat to public health and the natural ecosystem. A high degree of industrialization and urbanization has substantially enhanced the contamination of soil as well as aquatic environment through discharge of industrial wastewater and domestic wastes. Wastes from various industries mainly produce the hazardous pollutants. The production of these wastes will continue as long as human civilization persists. The remediation of hazardous pollutants has received attention since early 1980s mainly due to its toxicity and infectious nature. More than 30 industries have been identified by Ministry of Environment and Forest, Government of India, creating environmental pollution, out of which 16 industries *viz.* electroplating, dye and dye intermediates, textile, pulp and paper, dairy, paints, resins, pharmaceuticals, marble and slurry, tannery, fertilizer, battery manufacturing, pesticide formation, rubber processing, packaging materials etc. have been recognized as the most polluting industries.

Conventional technologies such as high–temperature incineration and various types of chemical decomposition (*e.g.*, base–catalyzed dechlorination, UVoxidation) have been used for the removal of pollutants. These technologies can be effective towards reducing the levels of contamination, but have several drawbacks, principally because of their technological complexity, the cost for small–scale application, and the lack of public acceptance, especially for incineration that may increase the exposure to contaminants for both the workers at the site and nearby residents.

Bioremediation is the use of living organisms, primarily microorganisms, to degrade the environmental contaminants into less toxic forms. It uses naturally occurring bacteria and fungi or plants to degrade or detoxify substances hazardous to human health and/or the environment. The microorganisms may be indigenous to a contaminated area or they may be isolated from elsewhere and brought to the contaminated site. Contaminant compounds are transformed by living organisms through reactions that take place as a part of their metabolic processes.

Yeasts are a heterogenous group of fungi. They grow in a conspicuous unicellular form and reproduce by fission, budding, or a combination of both. True yeasts reproduce sexually, developing ascospores or basidiospores under favourable conditions. The majority of ascomycetous and basidiomycetous yeasts isolated under laboratory conditions may not be identified because most of them are heterothallic. In many cases, one of the mating types is isolated and therefore, no asci or basidia are produced.

Yeasts have long been associated with human beings and being used as the most useful eukaryotic microorganisms for biological research. The importance of yeasts is understood by our daily consumption of bread and fermented beverages. Recent advances in biotechnology have increased our reliance on yeasts for production of pharmaceuticals and bulk biochemicals. Furthermore, clinically important yeasts also take important place, as the number of immunosuppressed patients are increasing. Biologists are continuing to discover the importance of yeasts in the ecosystem and their applications in biocontrol of plant pests. Therefore, almost all the areas of science and technology need rapid and accurate identification of yeasts. With all the historical background, commercial applications and applications to human diseases and research, yeasts did not receive universal recognition over the years as bioremediation agents like fungi and bacteria. This publication intends to highlight the importance of yeasts towards remediation of various pollutants from environment.

The yeast biomass has been successfully used as agents for removal of metal ions *viz.* Ag, Au, Cd, Co, Cr, Cu, Ni, Pb, U, Th and Zn from aqueous solution. Yeasts of different genera *viz. Saccharomyces, Candida, Pichia* etc. have shown efficiencies as biosorbents for the removal of heavy metal ions. Most of the yeasts can sequester wide range of metal ions or be strictly specific in respect of only one metal ion. *S. cerevisiae* as biosorbent is of special interest (Podgorskii *et al.*, 2004). A number of literatures have proved that *S. cerevisiae* can remove toxic metals, recover precious metals and clean radionuclides from aqueous solutions to various extents. The potentiality of *S. cerevisiae* as biosorbent in metal biosorption was depicted in detail by Wang and Chen (2006).

Over the past decade, many microorganisms have been reported for their abilities to remediate a variety of structurally diverse synthetic dyes. One of the most ubiquitous biomass types available for bioremediation of textile dyes at lower pH values is yeast. Compared to bacteria and filamentous fungi, yeasts exhibit attractive features. Yeasts are inexpensive, readily available source of biomass. Though not as fast as bacteria, yeast can grow faster than filamentous fungi, and like them, they have the ability to resist unfavourable environments (Yu and Wen, 2005; Gonen and Aksu, 2009a, b). Yeast cells retain their ability to accumulate a broad range of textile dyes to varying degrees under a wide range of external conditions (Donmez, 2002; Aksu, 2003).

Usage of diesel oil in passenger vehicles has been tremendously increased which creates an air pollution since diesel engines emit complex mixture of air pollutants composed of gaseous and solid particles. The visible emissions in diesel exhaust known as particulate matter contain carbon particles, harmful gases and 40 other cancer causing substances. Inhalation of these toxic substances may lead to lung cancer, premature death and other health problems (CARB, 2008). The most susceptible persons are children whose lungs are still developing and elderly who are having serious health problems. Yeasts have also been reported as potential agents for remediation of diesel oil. Two yeast strains *viz. Saccharomyces cerevisiae* and *Candida albicans* isolated from polluted lagoon water were found to grow effectively utilizing crude oil and diesel oil as sole sources of carbon and energy (Ilori *et al.*, 2008). The yeast *Trichosporon asahii*, isolated from petroleum–contaminated soil in India was reported as an efficient degrader of diesel oil (95%) over a period of 10 days (Chandran and Das, 2010).

For many years, we have enjoyed the benefits of using pesticides *viz.* lindane, DDT, DDD, endosulfan, kelthane, chlorobenzilate, chloropropylate, methoxychlor, aldrin, dieldrin, heptachlor, isodrin, isobenzan, endrin, chlordane, toxaphene, mirex, and kepone to control weeds, insects, pathogenic fungi, parasites and rodent pests. But apart from the beneficial role, the use of pesticides also has some drawbacks, such as potential toxicity to humans and other animals. Lindane or gamma isomer of hexachlorocyclohexane (γ–HCH) is a highly chlorinated, recalcitrant organochlorine pesticide (OCP), extensively used for the control of agricultural and medical pests (Polti *et al.*, 2014). There are reports on bioremediation of lindane using yeast species as potential candidate for bioremediation of lindane (Salam *et al.*, 2013; Salam and Das, 2013 a,b,c, Salam and Das, 2014a). Recently, a novel approach on lindane degradation in aqueous medium using a bio–nano hybrid system composed of yeast *Candida* sp. VITJzN04 and nano scale zinc oxide (n–ZnO) has been reported (Salam and Das, 2015).

The emerging pollutants pose potential hazards to humans and aquatic organisms (Bolong *et al.*, 2009). Caffeine, 1,3,7–trimethylxanthine, an alkaloid has been considered as an emerging pollutant (Rivas *et al.*, 2012). Wastewaters containing caffeine are discharged to the surrounding water bodies and cause environmental pollution and health hazards. Excessive consumption of caffeine through beverages is associated with a number of health problems (Babu *et al.*, 2005). Intake of caffeine in large amounts over extended period of time can lead to a condition known as caffeinism which causes unpleasant physical and mental conditions including nervousness, irritability, anxiety, muscle twitching (hyperreflexia), insomnia, headache, respiratory alkalosis, and heart palpitation.

Reports are available on the toxic effects of caffeine on women, animals and plants (Pincheira *et al.*, 2003; Meyer *et al.*, 2004). Though caffeine is considered to be toxic for many microorganisms, some microorganisms have the ability to grow on caffeine utilizing the same as a source of energy. The catabolic routes, however, are different in different groups of microbes reflecting the complexity in microbial caffeine degradation (Dash and Gummadi, 2006b). Caffeine degrading ability has been reported in various yeast species isolated from the soil of coffee plantation areas or from coffee dumps (Lakshmi and Das, 2013a).

Pollution from pharmaceuticals in the aquatic environment is now recognized as an environmental concern in many countries. Antibiotics have been considered as emerging pollutants over the past few years (Homen and Santos, 2011). Pharmaceutical industries involved in the production of antibiotics discharge their wastes openly which contain some quantity of active compounds which are toxic in nature. Globally, it is known to be present at a concentration ranging from a few micrograms to kilograms in water and soils posing a threat to the environment (Hamscher *et al.*, 2002). Various physico–chemical methods have been used for the treatment of pharmaceutical wastewaters which are of limited applicability because of limitations such as inefficiency of remediating high strength wastewater, high operating cost, huge labour requirement, high equipment cost, intervention of toxic by–products etc. (Homem and Santos, 2011). Thus, one way to solve these problems may be to use bioremediation technology involving microorganisms which are attracting increasing attention nowadays (Okoh, 2006). There are reports on the use of yeasts for the remediation of cephalosporin antibiotics from aqueous environment (Selvi and Das, 2015; Selvi and Das, 2014 a,b; Selvi *et al.*, 2014).

This book is intended to highlight the present status, feasibility and the significance of yeast mediated bioremediation to the academicians, students, researchers, environmentalists, industrialists, engineers and public who are enthusiastic to know about the utility of yeasts as novel microorganisms. I sincerely hope that this publication will definitely provide insights into the research frontier.

2

Yeast—A Novel Microorganism

Yeasts are single–celled microorganisms that are classified, along with molds and mushrooms, as members of the Kingdom Fungi. They are diverse and classified into two separate phyla, *Ascomycota* or sac fungi and *Basidiomycota* or higher fungi that together form the subkingdom *Dikarya*. Budding yeasts, also referred as 'true yeasts' are the members of the phylum *Ascomycota* and the order *Saccharomycetales*. Such classifications are based on characteristics of the cell, ascospore, and colony, as well as cellular physiology. Although yeasts are single–celled organisms, they possess a cellular organization similar to that of higher organisms, including humans. Specifically, their genetic content is contained within a nucleus. This classifies them as eukaryotic organisms, unlike their single–celled counterparts, bacteria, which do not have a nucleus and are considered prokaryotes. Some species with yeast forms may become multicellular through the formation of strings of connected budding cells known as pseudohyphae, or false hyphae, as seen in most of the molds. The size of yeast varies greatly depending on the species, typically measuring 3–4 µm in diameter, although some yeast species can reach the size over 40 µm. Most yeasts reproduce asexually by mitosis, and many do so by an asymmetric division process called budding.

HISTORICAL BACKGROUND

Yeast can be considered as oldest microorganism. Man started using yeast before the development of a written language. Egyptians started using yeast to produce alcoholic beverages through fermentation and to leaven bread over 5000 years ago. It is believed that these early fermentation systems for alcohol production and bread making were formed by natural microbial contaminants of flour, other milled grains and from fruit or other juices containing sugar. Wild yeasts and lactic acid bacteria were the microbial flora which was found to be associated with cultivated grains and fruits. For hundreds of years, it was traditional for bakers to obtain the yeast to leaven their bread and by–products of brewing and wine making. As a result, these early bakers selected yeasts as important industrial microorganisms. Based on the pioneering scientific work of Louis Pasteur, yeast was identified as a living organism and the agent responsible for alcoholic fermentation and dough leavening in the late 1860's. Shortly after these discoveries, it became possible to isolate yeast in pure culture form. With the knowledge of yeast as industrially important living organism, commercial production of baker's yeast was started around the turn of the 20th century. Since that time, bakers yeast manufacturers and environmental scientists have started working to find and produce pure strains of yeast to meet the specialized needs in various directions.

MORPHOTYPE

Syncytium is the first morphotype of yeast which is defined as one cytoplasm containing many nuclei. This morphotype is not very common in yeasts. Yeast species are able to form hyphal syncytia (true septate hyphae), also called filaments, and represented by the strictly hyphal species *Ashbya gossypii* from the genus *Eremothecium*. In hyphae of *A. gossypii*, the replicative cycle of individual nuclei is truly uncoupled from cell growth and the processes in the cytoplasm. In fact, each nucleus is able to follow its own replicative cycle, irrespective of the replicative stage of neighboring nuclei (Gladfelter, 2006).

The second morphotype is the 'normal' one nucleus–one cell arrangement, where nuclear division and cell division are tightly coupled via the cell cycle machinery. This coupling is evident during purely vegetative replication cycles of yeast cells. For many species, this normal cell division occurs also in the first cell division after mating and zygote formation.

The third mophotype is endobudding which occurs in other species where mating may be immediately followed by meiosis and spore formation. Endobudding is a process comparable to endodyodeny or endopolygeny in many protists such as *Toxoplasma gondii*. Here, cell replication takes place by the assembly of daughter cells inside the mother cells from nuclei that have been generated in one to several preceding nuclear divisions. This type of cell division involves a break of topology whereby each nucleus organizes its separation from the common cytoplasm by assembling a new plasma membrane that is discontinuous from the syncytial plasma membrane. In the yeasts, endobudding appears to occur only when linked to spore formation, which involves the concomitant assembly of a spore wall. Although this form of cell division is prevalent for the yeasts, no yeast species uses consecutive cycles of endobudding as a mode of reproduction (Knop, 2011).

HABITAT

Yeasts are widely dispersed in nature with a wide variety of habitats. They are commonly found on plant leaves, flowers, and fruits. They are also found on the surface of the skin and in the intestinal tracts of warm–blooded animals, where they may live symbiotically or as parasites. Some yeast species are found in association with soil and insects (Slávikoová and Vadkertiová, 2003; Suh *et al.*, 2005). They are also present in the gut flora of mammals (Martini, 1992). Yeasts, including *Candida albicans*, *Rhodotorula rubra*, *Torulopsis*, *Trichosporon cutaneum* and *Saccharomyces cerevisiae* may be found living in between people's toes as part of their skin flora (Oyeka and Ugwu, 2002). Most of the members of *Candida* species are found to colonise in the nectarines of flowers and honey stomachs of bees. *Dekkera intermedia* and *Candida blankii* were found to be the common species in honey stomachs and flower nectaries (Sandhu *et al.*, 1985).

Deep–sea environments also host an array of yeasts (Kutty and Philip, 2008). Eight yeast species were isolated from different hypersaline habitats worldwide, as well as from the Dead Sea, Enriquillo Lake (Dominican Republic) and the Great Salt Lake (Utah) (Butinar *et al.*, 2005). Among the isolates obtained from hypersaline waters, *Pichia guilliermondii, Debaryomyces hansenii, Yarrowia lipolytica* and *Candida parapsilosis* are known contaminants of low water activity food, whereas *Rhodosporidium sphaerocarpum, R. babjevae, Rhodotorula laryngis, Trichosporon mucoides*, and a new species resembling *C. glabrata* were not known for their halotolerance and were identified for the first time in hypersaline habitats. Moreover, the ascomycetous yeast *Metschnikowia bicuspidata*, known to be a parasite of the brine shrimp, was isolated as a free–living form from the Great Salt Lake brine. Two new species provisionally named *C. atmosphaerica*—like and *P. philogaea* like were discovered in water rich in magnesium chloride from the La Trinitat salterns (Spain). Hatha *et al.* (2013) isolated 34 yeast species from Kongsfjord at Ny Alseund region of Norwegion Arctic during Indian Arctic summer expedition of 2009. *Cryptococcus* was the predominant genus followed by *Trichosporon* and *Rhodotorula*. 82% of the yeast isolates were oxidative in nature. All the isolates used nitrate as nitrogen source except *Filobasidium*. Wide tolerance of low temperature ranging from 4°C to 20°C for the psychotrophic yeast isolates was observed from this study. Arthur and Watson (1976) reported that *Leucosporidium frigidum, Saccharomyces telluris* and *Candida slooffi* could grow at the range of –2 to 20°C, 5 to 35°C and 28 to 45°C respectively.

Importance

Yeasts are being used in the various fields of biotechnology because of their useful physiological properties. The yeasts *viz. Saccharomyces cerevisiae* and *Schizosaccharomyces pombe* have been considered as industrially important conventional yeast species being used in food and beverage industries for millennia. Other non conventional yeasts *viz. Schwanniomyces occidentalis, Kluyveromyces lactis, Pichia pastoris, Pichia methanolica* (*Pichia pinus*), *Pichia angusta* (*Hansenula polymorpha*), *Pichia guilliermondii, Yarrowia lipolytica, Candida maltose, Arxula adeninivorans*, and *Trichosporon* sp. are of special importance. The useful functions of yeasts include: *Arxula adeninivorans* assimilates adenine and other purines. *S. accidentalis* metabolizes starch completely and *K. lacis* utilizes lactose. *Pichia pastoris, Pichia methanolica* (*Pichia pinus*), and *Pichia angusta* (*Hansenula polymorpha*) metabolize methanol and are especially useful as vehicles for production of heterologous proteins. *Pichia guilliermondii* produces riboflavin. *Yarrowia lipolytica* and *Candida maltosa* metabolise hydrocarbons, fatty acids, other lipids and related compounds. *Y. lipolytica* produces citric and other organic acids. *Trichosporon* spp. may be useful as microbial sensors for determining biological oxygen demand (BOD) of waste waters, carbohydrate analysis, amperometric determination of ammonium ions, and phenol determination (Reiser *et al.*, 1994).

Many types of yeasts are used for making foods like bread production, beer fermentation, wine fermentation and for xylitol production (Rao *et al.*, 2004). *Saccharomyces* yeasts have been genetically engineered to ferment xylose, one of the major fermentable sugars present in cellulosic biomasses, such as agriculture residues, paper wastes, and wood chips (Brat *et al.*, 2009). Therefore, ethanol could be efficiently produced from inexpensive feedstocks, making cellulosic ethanol fuel, a lower priced alternative to gasoline fuels (Madhavan *et al.*, 2012). Yeasts are considered as most widely used model organisms for genetics and cell biology (Botstein and Fink, 2011). Yeast in symbiotic association with acetic acid bacteria is used in the preparation of *kombucha*, a fermented sweetened tea. *Kefir* and *kumis* are made by fermenting milk with yeast and bacteria (Leite *et al.*, 2013). The yeast *S. boulardii* is being used as probiotic supplement to maintain and restore the natural flora in the gastrointestinal tract and has been used to reduce (i) the incidence of HIV/AIDS–associated diarrheas (McFarland, 2010) (ii) the symptoms of acute diarrhoea (Dinleyici *et al.*, 2012), (iii) the chance of infection caused by *Clostridium difficile* (Johnson *et al.*, 2012) and to reduce bowel movements in diarrhoea–predominant IBS patients (Dai *et al.*, 2013).

The potential applications of yeasts in the field of bioremediation have been reported. The yeast *Yarrowia lipolytica* could tolerate high concentrations of salt and heavy metals (Bankar *et al.*, 2009) and was investigated for its potential as a heavy metal biosorbent (Bankar *et al.*, 2009). *Saccharomyces cerevisiae* showed potentiality to bioremediate toxic pollutants like arsenic from industrial effluent (Soares and Soares, 2012). The yeast, *Pichia fermentans* MTCC 189 to remediate synthetic dyes (Das *et al.*, 2010), *Candida* sp. VITGBN1 and *Cryptococcus* sp. VITGBN2 to remediate heavy metals (Das *et al.*, 2012), *Candida tropicalis* to degrade diesel (Chandran and Das, 2011), *Rhodotorula* sp. VITJzN03 and *Candida* sp. VITJzN04 to remediate organochlorine pesticide lindane (Salam *et al.*, 2013; Salam and Das, 2014), *Trichosporin asahii* to remediate caffeine (Lakshmi and Das, 2013) and *Ustilago sparsa* SMN03 to degrade cefdinir, a β–lactum antibiotic from aqueous environment (Selvi *et al.*, 2014) have also been reported.

ISOLATION OF YEAST

All initially isolated yeasts may be contaminated in mixed culture. The purity of each yeast isolate has to be confirmed by direct mounts and subsequent streaking for colony isolation. Pure cultures are necessary for assimilation experiments. Streaking subcultures for spatial isolation is adequate in most cases. To obtain pure culture of yeast, following techniques can be used.

Isolation Techniques from Mixed Cultures

Colonies of mixed yeasts

- Suspend a portion of each suspected colony type in a tube of sterile distilled water
- Streak a loopful of suspension onto a SAB agar plate
- Incubate the SAB plate at 30°C for 48 hours
- Check for purity. If the yeast is not pure, it must be restreaked

Mould contamination

1. Colony isolation on yeast malt (YM) agar
   - Suspend a small portion of each suspected colony type in a tube of sterile distilled water
   - Streak a loopful of suspension for colony isolation onto a plate of YM agar
   - Incubate the YM plate at 30°C for 4–6 days
   - Check the purity. If not pure, further step is necessary (Step 2)
2. Colony isolation on yeast malt broth
   - Transfer a small portion of the yeast mold isolate to a tube containing 10 ml of YM broth
   - Incubate the YM broth at 30°C for 48 hours. Remove a small portion of the sediment with a sterile capillary pipette by slipping along the edge of the tube to the bottom without disturbing the mycelial pellicle
   - Streak the sediment for colony isolation onto a plate of YM agar
   - Incubate at 30°C for 4–7 days and then prepare a direct mount. If the yeast is not pure, further step is necessary (Step 3)
3. Colony isolation in shake culture
   - Transfer a small portion of the yeast mold isolate to a 250 ml Erlenmeyer flask containing approximately 100 ml of YM broth
   - Place the flask on rotary shaker and incubate at 30°C for 4–6 days
   - Small portion of the sediment has to be removed with a sterile capillary pipette
   - Ensure that the balls of the mycelium are not removed
   - Streak the sediment for colony isolation onto agar late of YM agar
   - Incubate at 30°C for 4–7 days and then prepare a direct mount using a small portion of each yeast colony to be identified.

Bacterial contamination

1. Colony isolation on SAB agar
   - A small portion of yeast to be decontaminated in sterile distilled water
   - Streak a loopful of the suspension for colony isolation onto a plate of SAB agar
   - Incubate at 30°C for 48 hours
   - Examine for isolated colonies. Purity of the colonies to be verified
   - If colonies are not pure, further step is necessary (Step 2)
2. Colony isolation on SAB agar plus chloramphenicol, SAB plus penicillin and streptomycin or BHI plus 10% blood gentamicin and chloramphenicol
   - Suspend a small portion of the yeast to be decontaminated in sterile distilled water
   - Streak a loopful of the suspension for colony isolation onto one of the above media
   - Incubate at 30°C for 48 hours. Check purity. If colonies are not pure, further step is necessary (Step 3)
3. Acidification of SAB broth
   - Suspend a small portion of the yeast to be decontaminated in sterile distilled water
   - To each of 4 tubes containing 10 ml of SAB broth, add 1 drop of 1N HCl to the first tube, 2 drops to the second tube, 3 drops to the third and four drops to the fourth tube
   - Add 0.5 ml of the contaminated yeast suspension to each tube
   - Incubate at 30°C for 24 hours
   - Check purity

IDENTIFICATION OF YEAST

Identification of yeast is mainly based on morphological and biochemical criteria. Morphological identification is needed for the establishment of genera and biochemical studies are necessary for the differentiation of the various species. The principal criteria for identifying the yeasts include (a) cultural characteristics (b) asexual structures (c) sexual structures and (d) physiological studies. The cultural characteristics include colony colour, shape and texture. The asexual structures include shape and size of the cells, unipolar/ bipolar/multipolar budding, absence or presence of conidia (arthroconidia, ballistoconidia, blastoconidia), endoconidia, presence of hyphae/pseudohyphae and, sporangia. Sexual structures include number, size and shape of ascospores and basidiospores. Physiological studies involve various tests *viz.* assimilation, cycloheximide resistance, fermentation, nitrogen utilization, urea hydrolysis and growth under different temperature.

Identification of Yeasts through Biochemical Analysis

The yeasts can be identified to the species level through biochemical analysis using VITEK 2 compact yeast card reader with the software version: 03.01.

Principle behind VITEK 2 Compact Yeast Card Reader

The VITEK 2 is an automated microbiology system utilizing growth–based technology. The system is available in three formats (VITEK 2 compact, VITEK 2, and VITEK 2 XL) that differ in increasing levels of capacity and automation. All three systems accommodate the same colorimetric reagent cards that are incubated and interpreted automatically.

Reagent Cards

The reagent cards have 64 wells, each contain an individual test substrate. Substrates measure various metabolic activities such as acidification, alkalinisation, enzyme hydrolysis and growth in the presence of inhibitory substances. An optically clear film present on both sides of the card allows for the appropriate level of oxygen transmission while maintaining a sealed vessel that prevents contact with the organism–substrate mixtures. Each card has a pre–inserted transfer tube used for inoculation. Cards have bar codes that contain information on product type, lot number, expiration date, and a unique identifier that can be linked to the sample either before or after loading the card onto the system.

Culture Requirements

The on–line product information contains a culture requirements table that lists parameters for appropriate culture and inoculum preparation. These parameters include acceptable culture media, culture age, incubation conditions and inoculum turbidity.

Suspension Preparation

A sterile swab or applicator stick is used to transfer a sufficient number of colonies of a pure culture and to suspend the microorganism in 3 mL of sterile saline (aqueous 0.45 to 0.50% NaCl, pH 4.5 to 7.0) in a 12 × 75 mm clear plastic (polystyrene) test tube. The turbidity is adjusted accordingly and measured using a turbidity meter called the DensiChek™. McFarland turbidity for yeast species exists in the range of 1.80–2.20.

Inoculation

Identification cards are inoculated with microorganism suspensions using an integrated vacuum apparatus. A test tube containing the microorganism suspension is placed into a special rack (cassette) and the identification card is placed in the neighbouring slot while inserting the transfer tube into the corresponding suspension tube. The cassette can accommodate up to 10 tests. The filled cassette is placed manually into a vacuum chamber. After the vacuum is applied and air is re–introduced into the station, the organism suspension is forced through the transfer tube into micro–channels that fill all the test wells.

Card Sealing and Incubation

Inoculated cards are passed by a mechanism, which cuts off the transfer tube and seals the card prior to loading into the carousel incubator. The carousel incubator can accommodate up to 30 or up to 60 cards. All card types are incubated on–line at 35.5 ± 1.0°C. Each card is removed from the carousel incubator once every 15 min, transported to the optical system for reaction readings and then return to the incubator until the next read time. Data are collected at 15 min intervals during the entire incubation period.

Optical System

A transmittance optical system allows interpretation of test reactions using different wavelengths in the visible spectrum. During incubation, each test reaction is read every 15 min to measure either turbidity or coloured products of substrate metabolism. In addition, a special algorithm is used to eliminate false readings due to small bubbles that may be present.

Test Reactions

Calculations are performed on raw data and compared to thresholds to determine reactions for each test. On the VITEK 2 Compact, test reaction results appear as "+", "–", "(–)" or "(+)". Reactions that appear in parentheses are indicative of weak reactions that are too close to the test threshold.

Database Development

The databases of the VITEK 2 identification products are constructed with large strain sets of well–characterized microorganisms tested under various culture conditions. The yeast strains may be derived from a variety of clinical and industrial sources as well as from public (*e.g.*, ATCC) and university culture collections.

Analytical Techniques

Test data from an unknown organism needs to be compared to the respective database to determine a quantitative value for proximity to each of the database taxa. Each of the composite values is compared to the others to determine if the data are sufficiently unique or close to one or more of the other database taxa. If a unique identification pattern is not recognized, a list of possible organisms is given or the strain will be determined to be outside the scope of the database.

Identification Levels

An unknown biopattern needs to be compared to the database of reactions for each taxon and a numerical probability calculation will be performed. Various qualitative levels of identification are assigned based on the numerical probability calculation. The different levels and associated information are shown in Table 2.1.

Table 2.1: Identification levels.

ID message	*Choices*	*% probability*	*Comments*
Excellent	1	96 to 99	N/A
Very good	1	93 to 95	N/A
Good	1	89 to 92	N/A
Acceptable	1	85 to 88	N/A
Low	2 to 3	Sum of choices = 100 After resolution to one Choice, percentage probability reflects the number associated with the choices	2 to 3 taxa exhibit same biopattern Separate by supplemental test

Yeast Card (YST)

The YST card is used for the automated identification of 49 taxa of the most significant yeasts and yeast–like organisms. The YST identification card is based on established biochemical methods and newly developed substrates. There are 46 biochemical tests measuring carbon source utilization, enzymatic activities and resistance. Final identification results are available in approximately 18 hours.

The list of test substrates is shown in Table 2.2. In a recent multi–site study, the performance of the VITEK 2 YST are evaluated using 623 isolates of both commonly and rarely observed species of yeast and yeast–like organisms. The reference identification is determined with api 20C AUX identification kits. Overall, the VITEK 2 YST correctly identifies 98.9% of the isolates, including 11.7% low discrimination with the correct species listed. Misidentifications occur at 0.6% and no identifications occur at 0.5%.

Table 2.2: Test substrates on YST card.

Test	*Mneumonic*
L–lysine arylamidase	LysA
L–Malate assimilation	IMLTa
Leucine arylamidase	LeuA
Arginine GP	ARG
Erythritol assimilation	ERYa
Glycerol assimilation	GLYa
Tyrosine arylamidase	TyrA
Beta–N–Acetylglucosaminidase	BNAG
Arbutin assimilation	ARBa
Amygdalin assimilation	AMYa
D–Galactose assimilation	dGALa
Gentiobiose assimilation	GENa
D–Glucose assimilation	dGLUa
Lactose assimilation	LACa
Methyl–A–D–Glucopyranoside assimilation	MAdGa
D–Cellobiose assimilation	dCELa
Gamma Glutamyl Transferase	GGT
D–Maltose assimilation	dMALa
D–Raffinose assimilation	dRFAa
PNP–N–Acetyl–BD–Galactosaminidase 1	NAGA1
D–Mannose assimilation	dMNEa
D–Melibiose assimilation	dMELa
D–Melezitose assimilation	dMLZa
L–Sorbose assimilation	ISBEa
L–Rhamnose assimilation	IRHAa
Xylitol assimilation	XLTa
D–Sorbitol assimilation	dSORa

Table 2.2: (*Contd...*)

Table 2.2: *(Contd...)*

Test	*Mneumonic*
Saccharose/Sucrose assimilation	SACa
Urease	URE
Alpha Glucosidase	AGLU
D–Turanose assimilation	dTURa
D–Trehalose assimilation	dTREa
Nitrate assimilation	NO3a
L–Arabinose assimilation	IARAa
D–Galacturonate assimilation	dGATa
Esculin hydrolysis	ESC
L– Glutamate assimilation	IGLTa
D– Xylose assimilation	dXYLa
DL– Lactate assimilation	LATa
Acetate assimilation	ACEa
Citrate (sodium) assimilation	CITa
Glucuronate assimilation	GRTas
L–Proline assimilation	IPROa
2– Keto– D– Gluconate assimilation	2Kga
D– Gluconate assimilation	dGNTa

Molecular Identification of Yeast Strain

For molecular identification of yeast strain, genomic DNA extraction, partial and complete sequences of 18S rDNA, ITS1–5.8S rDNA–ITS2 and 28S rDNA needs to be done as follows:

Genomic DNA Extraction

Genomic DNA extraction can be performed following the method of Cheng and Jiang (2006).

- Yeast cells to be grown on YEPD broth for 48 h at 30°C and harvested by centrifugation at 5000 X *g* for 10 min.
- The cell pellets need to be resuspended 500 µL of sterile deionised water.
- Addition of 50 mg of 425–600 µm size–fractionated glass beads to cell suspension and then 100 µL of Tris–saturated phenol

(pH 8.0), followed by vortex–mixing step of 120–200s to the lyse cells.

- Centrifugation of the sample at 13000 X *g* for 5 min at 4°C. Transfer of upper aqueous phase (160 μL) to a clean 1.5 mL tube.
- Addition of 40 μL TE buffer (10 mM Tris–HCl, pH 8.0; 1mM EDTA, pH 8.0) to make 200 μL, mix with 100 μL chloroform and centrifuge for 5 min at 13000 X *g* at 4°C. Lysate to be purified by chloroform extraction until a white interface is no longer present.
- Transfer of 160 μL of upper aqueous phase to a clean 1.5 mL tube. Addition of 40 μL TE and 5 μL RNase (at 10 mg/mL) and incubate at 37°C for 10 min to digest RNA.
- Addition of chloroform (100 μL) to the tube, mix well and centrifuge for 5 min at 13000 X *g* at 4°C. Transfer of upper aqueous phase (150 μL) to a clean 1.5 mL tube. The aqueous phase contains purified DNA and directly can be used for the experiments or store at –20°C. The quality and integrity of the DNA can be verified by electrophoresis in 1% agarose gel and visualized by ethidium bromide staining.

PCR Amplification of the 18S–ITS1–5.8S–ITS2–28S rDNA

The amplification of the DNA fragment coding for the 18S–ITS1–5.8S–ITS2–28S rRNA (Fig. 2.1) can be done using the synthetic primers: UL18F:5′–TGTACACACCGCCCGTC–3′ and UL28R:5′–ATCGCC AGTTCTGCTTAC–3′. PCR amplification are to be performed with a volume of 50 μL. 2 μL of DNA template needs to be added to the PCR master mixture, which consist of 5 μL of 10 X PCR buffer (with 15 mM $MgCl_2$), 4 μL of dNTPs (0.1 mM each dNTP), 0.8 μL of each primer (40 pmol of each primer) and 0.4 μL (2.0 U) of Taq polymerase with the remaining volume consisting of distilled water. PCR can be performed in Eppendorf mastercycler, with an initial denaturation at 94°C for 4 min followed by 30 cycles of denaturation at 94°C for 30s, annealing at 55°C for 30 s and extension at 72°C for 1 min. A final extension period of 5 min at 72°C needs to be carried out at the end of the 30 cycles. PCR products need to be separated on 2% agarose gel. DNA bands need to be purified using a QIA quick gel extraction kit and the conserved regions need to be sequenced by using BigDye® Terminator v3.1 Cycle Sequencing Kit, following Sanger sequencing method on ABI 3730 XL analyzer.

UL18F:5′–TGTACACACCGCCCGTC–3′, UL28R: 5′ATCGCCA GTTCTGCTTAC–3′; ITS3:5′–GCATCGATG AAGAAC GCAGC–3′ (universal primer) and UL620R:5′–TGGTCCGT GTTTCAAGA–3′ (Acme ProGen) primer sequences can be used for sequencing the conserved regions.

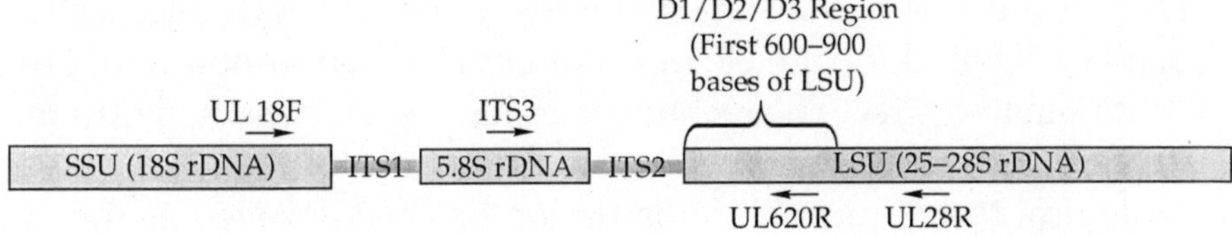

Fig. 2.1: Schematic representation of the yeast ribosomal genes containing the primer target areas (Martin and Rygiewicz, 2005).

Phylogenetic Analysis

The partial sequences obtained need to be subjected to the BLASTN search version 2.2.3 (Altschul *et al.*, 1990) in the National Centre for Biotechnology Information (NCBI) database (http://www.ncbi.nih.gov). A collection of taxonomically related sequences obtained from this database can be used to perform a multiple sequence alignments using the CLUSTAL X program. Phylogenetic and molecular evolutionary analyses can be conducted by MEGA 5.1 software with Kimura 2–parameter model (Tamura *et al.*, 2011). Phylogenetic trees can be constructed using the neighbor–joining method, and 1000 bootstrap replications can be assessed to support internal branches (Hillis and Bull, 1993).

3

Microbial Bioremediation

During the last two decades, a wide variety of technologies have been developed for the clean up of environmental pollutants from contaminated sites. The technologies have been classified into three categories *viz.* physical, chemical and biological. Among the biological techniques, bioremediation has been proved as the most promising one (Saval, 2003).

DEFINITION

Bioremediation is a technique which involves the use of microorganisms for the degradation of hazardous chemicals from soil, sediments and water. Often the microorganisms metabolize the chemicals to produce carbon dioxide or methane, water and biomass (Harms *et al.*, 2011; Silar *et al.*, 2011). Alternatively, the contaminants may be enzymatically transformed to metabolites that are less toxic or innocuous (Iwamoto and Nasu, 2001).

BIOREMEDIATION APPROACHES

Bioremediation is an organic approach to the reclamation of waste materials at the site. This technology is now routinely used for the treatment of wastewater, soil, sludges, ground water and surface water contaminated with various pollutants *viz.* petroleum hydrocarbons, pesticides, heavy metals, synthetic dyes, antibiotics etc (Fig. 3.1).

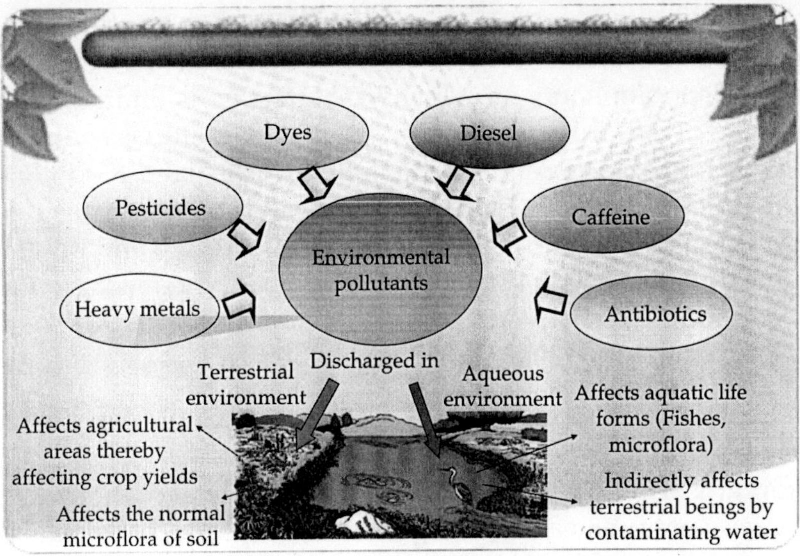

Fig. 3.1: Release of various contaminants causing pollution to the environment.

Detoxification and mineralization of the pollutants to biomass, production of CO_2 and water make it an attractive, environment–friendly, safe and cost effective technology alternative to conventional technologies. Naturally occurring bacteria and fungi or plants are used to degrade or detoxify substances hazardous to human health and/or the environment. The microorganisms may be indigenous to a contaminated area or they may be isolated from elsewhere and brought to the contaminated site. Contaminant compounds are transformed by living organisms through reactions that take place as a part of their metabolic processes. Biodegradation of a compound is often a result of the actions of multiple organisms. When microorganisms are imported to a contaminated site to enhance degradation, the process is known as bioaugmentation. For bioremediation to be effective, microorganisms must enzymatically attack the pollutants and convert them to harmless products.

Microbial bioremediation offers the possibility to destroy or render harmless various contaminants using natural biological activity of microorganisms. As such, it uses relatively low–cost techniques, which generally have a high public acceptance and can often be carried out on site. It will not always be suitable, however, as the range of contaminants on which it is effective is limited, the time scales involved are relatively long, and the residual contaminant levels achievable may not always be appropriate. Although the methodologies employed are not technically complex, considerable experience and expertise may be required to design and implement a successful bioremediation program. Because bioremediation seems to be a good alternative to conventional clean–up technologies, it has been used at a number of sites worldwide.

There are at least five critical factors that should be considered when evaluating the use of bioremediation for site cleanup. These factors are: (a) Magnitude, toxicity, and mobility of contaminants (b) Proximity of human and environmental receptors, (c) Degradability of contaminants (d) Planned site use and (e) Ability to monitor properly. Different bioremediation techniques are employed depending on the degree of saturation and aeration of an area (Tarangini *et al.*, 2009; Qaiser *et al.*, 2009).

TYPES OF BIOREMEDIATION

The bioremediation treatment technologies can be categorized mainly into two types: *In situ* bioremediation and *Ex situ* bioremediation. *In situ* techniques are defined as those which are applied to the soil and groundwater at the site with minimal disturbance. *Ex situ* techniques are those which are applied to soil and groundwater at the site that has been removed from the site *via* excavation (soil) or pumping (water). *In situ* techniques include biosparging, bioventing and bioaugmenation whereas *Ex situ* techniques include landfarming, composting, biopiles, bioreactors etc.

In situ techniques are generally the most desirable options due to low cost and less disturbance since they provide the treatment in place avoiding excavation and transport of contaminants. This treatment is limited by the depth of the soil that can be effectively treated. In many soils, effective oxygen diffusion for desirable rates of bioremediation extend to a range of only a few cm to about 30 cm into the soil, although depths of 60 cm and greater have been effectively treated in some cases (Vidali, 2001). Biosparging involves the injection of air under pressure below the water table to increase groundwater oxygen concentrations and enhance the rate of biological degradation of contaminants by naturally occurring microorganisms. It increases the mixing in the saturated zone and thereby increases the contact between soil and groundwater. The ease and low cost of installation allows considerable flexibility in the design and construction of the system. Bioventing is the most common *in situ* treatment and involves supplying air and nutrients through wells to the contaminated soil for stimulating the indigenous bacteria. Bioventing employs low air flow rates and provides only the amount of oxygen necessary for the biodegradation while minimizing volatilization and release of contaminants to the atmosphere. It works for simple hydrocarbons and can be used where the contamination is deep under the surface. Bioaugmentation involves the addition of bacterial cultures to a contaminated medium. This method is frequently used in bioreactors. Two factors limit the use of added microbial cultures in a land treatment unit which include: (1) nonindigenous cultures rarely compete well enough with an indigenous population to develop and sustain useful population levels and (2) most soils with long–term exposure to biodegradable waste have indigenous microorganisms that are effective degraders if the land treatment unit is well managed.

Ex–situ techniques mainly involve the excavation or removal of contaminated soil from ground. Landfarming is a solid phase treatment for contaminated soil which is excavated and spread over a prepared bed and periodically tilled until pollutants are degraded. The goal is to stimulate indigenous biodegradative microorganisms and facilitate their aerobic degradation of contaminants. The practice is limited to the treatment of superficial 10–35 cm of soil. Since landfarming has the potential to reduce monitoring and maintenance costs, as well as clean–up liabilities, it has received much attention as a disposal alternative. Composting is an aerobic, thermophilic treatment process in which contaminated materials are mixed with nonhazardous organic materials such as manure or agricultural wastes. The presence of these organic materials supports the development of a rich microbial population. Biopiles are a hybrid of landfarming and composting. Essentially, engineered cells are constructed as aerated composted piles. Biopiles provide a favorable environment for indigenous aerobic and anaerobic microorganisms (Von Fahnestock *et al.*, 1998). Bioremediation in reactors involves the processing of contaminated solid material (soil, sediment, sludge) or water through an engineered containment system. Reactors are used to treat liquid or slurry or treatment of contaminated soil and water pumped up from a contaminated plume. A slurry bioreactor may be defined as a containment vessel and apparatus used to create a three–phase (solid, liquid, and gas) mixing condition to increase the bioremediation rate of soil bound and water–soluble pollutants as a water slurry of the contaminated soil and biomass (usually indigenous microorganisms) capable of degrading target contaminants.

On site bioremediation of polluted soil uses three types of bioreactors (1) static beds and heaps of soil that can be intermittently irrigated, aerated and mixed (2) rotating cylindrical and horizontal screw mixed moving bed systems that are continuously aerated and mixed and intermittently irrigated and (3) high and low aspect ratio bioreactors in which soil undergoing treatment is present as an aqueous slurry subjected to aeration and agitation (Jogdand, 2006).

FACTORS AFFECTING BIOREMEDIATION

Bioremediation is a complex process depending upon many factors. Physico–chemical factors *viz.* pH, nutrients, temperature, moisture and inhibitory substances or metabolites are the important factors that may affect bioremediation (Boopathy, 2000). The key elements in bioremediation process are the aerobic and anaerobic microorganisms. The microorganisms must be present in good numbers in bioremediation site. Their growth and activity must be stimulated. Biostimulation usually involves the addition of nutrients and oxygen to help indigenous microorganisms. These nutrients are the basic building blocks of life and allow microbes to create the necessary enzymes to break down the contaminants. All of them need nitrogen, phosphorous, and carbon. Carbon is the most basic element of living forms and is needed in greater quantities than other elements. In addition to hydrogen, oxygen and nitrogen, it constitutes about 95% of the weight of cells. Phosphorous and sulfur contribute with 70% of the remainders. The nutritional requirement of carbon to nitrogen ratio is 10:1, and carbon to phosphorous is 30:1.

Environmental factors *viz.* pH, temperature, and moisture affect the microbial growth and activity. Though microorganisms are isolated in extreme conditions, most of them grow under optimal conditions only. If the soil is too much acidic, it is possible to rinse the pH by adding lime. Temperature affects biochemical reactions rates. Above a certain temperature, the microbial cells do not survive. Plastic covering is applied to enhance solar warming in summer, late spring, and autumn if needed. All living organisms have water requirement. Therefore, irrigation is needed to achieve the optimal moisture level. The amount of available oxygen will determine whether the system is aerobic or anaerobic. Hydrocarbons are readily degraded under aerobic conditions, whereas chlorurate compounds are degraded only in anaerobic ones. To increase the oxygen amount in the soil, it is necessary to till or sparge air. In some cases, hydrogen peroxide or magnesium peroxide can be introduced in the environment. Soil structure is important which controls the effective delivery of air, water, and nutrients. Materials such as gypsum or organic matter are added to improve the soil structure. Low soil permeability can hinder the movement of water, nutrients, and oxygen; hence, soils with low permeability is not appropriate for *in situ* remediation (Vidali, 2001).

Microbial Interactions for Bioremediation

Genetic, biochemical and ecological abilities of microbes play an important role in remediation of pollutants. Microbial remediation under various conditions can be controlled through proper reactor designing. For successful bioremediation, essential characteristics of microbes include: (i) capacity to degrade pollutants, (ii) capable of utilizing pollutant as energy and carbon source, (iii) to synthesize the specific enzymes for the target compounds, (iv) presence of proper electron acceptor and donor system and (v) absence of toxic substances.

Microorganisms exhibit positive and negative interactions resulting growth, metabolism and survival which are essential for bioremediation. Positive interactions include commensalism, synergism and mutualism whereas negative interaction includes competition, parasitism and predation.

Mechanism of Bioremediation

The bioremediation mechanisms may be classified based on various criteria. According to the dependence on the cell's metabolism

- Metabolism dependent and metabolism independent
- According to the location where the pollutant removed from solution is found
    - Extra-cellular accumulation/precipitation
    - Cell surface sorption/precipitation and intracellular accumulation

Transport of the pollutant across the cell membrane of microbes yields intracellular accumulation, which is dependent on the cell's metabolism. This means that this kind of accumulation may take place only with viable cells. It is often associated with an active defense system of the microorganism, which reacts in the presence of toxic pollutant. In non metabolism dependent process, pollutant uptake occurs by physico–chemical interaction between the polluttant and the functional groups present on the microbial cell surface. This is based on physical adsorption, ion exchange and chemical sorption, which is not dependent on the cells' metabolism. Cell walls of microbial biomass, mainly composed of polysaccharides, proteins and lipids have abundant pollutant binding groups such as carboxyl, sulphate, phosphate and amino groups. The non–metabolism dependent process is relatively rapid and can be reversible (Prasad *et al.*, 2011). In the case of precipitation, the pollutant uptake may take place both in the solution and on the cell surface (Shroff and Vaidya, 2012). Further, it may be dependent on the cell's metabolism, if in the presence of toxic pollutant, the microorganism produces compounds that favor the precipitation process. Precipitation may not be dependent on the cells' metabolism, if it occurs after a chemical interaction between the pollutant and cell surface.

The chief ways by which remediation may be accomplished include biosorption, bioaccumulation and biodegradation (Fig. 3.2). Biosorption is a property of certain types of inactive, dead microbial biomass to bind contaminants from even very dilute solutions. Biosorption has emerged as an alternative and sustainable strategy for cleaning up water that has been contaminated with pollutants by anthropogenic activities and/or by natural processes, as it is effective, cheap and eco–friendly (Suazo–Madrid *et al.*, 2011). This technology has potential marketing advantages over other traditional wastewater treatment technologies as it is cheaper and environmental friendly, particularly when natural biomass is used (Chatterjee *et al.*, 2010; Zhang *et al.*, 2010). Mechanism of biosorption includes ion exchange, chelation and complexation. Several active functional groups of cell constituents like acetamide group of chitin, structural polysaccharides, amine, sufhydryl and carboxyl groups in protein, phosphate and hydroxyl groups in polysaccharide participate in biosorption.

Bioaccumulation is defined as the accumulation of pollutants by actively growing cells by metabolism and temperature–independent and metabolism–dependent mechanism steps (Saddetin and Donmez, 2006). Direct use of growing cells can avoid the need for separate biomass production processes (*e.g.*, cultivation, harvesting, drying, processing and storage prior to use) and can be very efficient under appropriate conditions such as moderate pollutant concentrations, solution pH, and salt concentrations that favour active growth of cells (Wang and Hu, 2008).

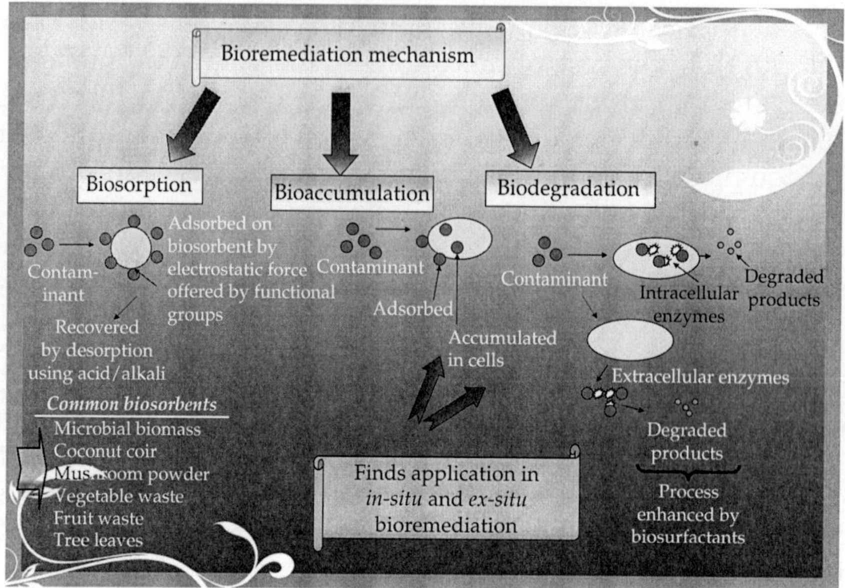

Fig. 3.2: Bioremediation mechanism.

Biodegradation is one of the processes involved in bioremediation which involves natural populations of microorganisms by which pollutants can be eliminated from the environment (Leahy and Colwell, 1990). Degradation of the organic type of contaminants by microbes leads to complete mineralization by releasing carbon dioxide. Moreover, it is cheaper than other remediation technologies (Ulrici, 2000). In this process, the parent molecule becomes detoxified or rendered harmless to life by enzymatic reactions. Presence of co–substrates might be required to support the energy yielding activity. Biotransformation or bioconversion makes them less toxic entities. Microbial biodegradation occurs due to the production of certain enzymes which acts as catalysts for transformation reactions induced by the energy yielding process. Energy is generated during breakdown of complex organic substances by heterotrophs, oxidation of inorganic compounds by autotrophs and photosynthetic reactions of phototrophs (Jogdand, 2004). Extracellular enzymes produced by the microbes break down the larger complex molecules and enable their entry into the cell. The enzymes which carry on different reactions in microbial cells are responsible for metabolic pathways forming a base for energy generation and cell synthesis (Rajendran and Gunasekaran, 2006).

ADVANTAGES AND LIMITATIONS OF BIOREMEDIATION

Advantages

- Bioremediation is a natural process and useful for the removal of wide variety of pollutants
- Microbes which are capable to degrade the contaminant increase in numbers when the contaminant is present. When the contaminant is degraded, the biodegradative population declines. The residues for the treatment are usually harmless products
- Many hazardous compounds can be transformed to harmless products and include carbon dioxide, water, and cell biomass
- Instead of transferring contaminants from one environmental medium to another, for example, from land to water or air, the complete destruction of target pollutants is possible
- Bioremediation can often be carried out on site, often without causing a major disruption of normal activities. This also eliminates the need to transport quantities of waste off site and the potential threats to human health and the environment that can arise during transportation
- Bioremediation can be less expensive compared to other technologies that are used for clean–up of hazardous materials
- It is therefore perceived by the public as an acceptable waste treatment process for contaminated sites

Limitations

- Bioremediation is limited to biodegradable compounds only. Not all compounds are susceptible to rapid and complete degradation

- The biodegradation products sometimes may be more persistent or toxic than the parent compound
- Biological processes are often highly specific. Important site factors include the presence of metabolically capable microbial populations, suitable environmental growth conditions, appropriate levels of nutrients and contaminant for success.
- It is difficult to extrapolate from bench and pilot–scale studies to full–scale field operations
- Sometimes longer time is needed than other treatment options, such as excavation and removal of soil or incineration
- There is a persistent need for demonstrations processes and their evaluation
- Comprehensive pollutant audits are required on waste streams and contaminated sites if effective biotreatment processes are to be developed
- Innovative biotreatment technologies should be responsive to public concern

4

Remediation of Heavy Metals Using Yeasts

Release of high concentration of heavy metals causes toxicity to the environment. Increased knowledge about the toxicological effects of heavy metals is well recognized and therefore, in recent years, removal of heavy metals has received a great deal of attention in the field of environmental pollution control and abatement (Fomina and Gadd, 2014). Heavy metals like cadmium, lead, copper, nickel, zinc etc. are usually found in many vital industries like ceramics, metal finishing, paper as well as petrochemicals. The treatment and disposal of heavy metal bearing wastes presents a unique challenge to the environmental scientists.

Conventional methods for heavy metal removal include reverse osmosis (Aljendeel, 2011), ion exchange (Shek *et al.*, 2009), electrofloatation (Casquaeria *et al.*, 2006), chemical precipitation (Ghosh *et al.*, 2011), ultrafiltration (Rahmanian, 2010) and adsorption (Katsou *et al.*, 2011). However, application of such methods has been always expensive and ineffective in terms of energy and chemical product consumption, especially at low concentrations of 1–100 mg/L. An alternate cost effective treatment strategy is required which will be eco–friendly.

Recent research indicates that microorganisms can accumulate high concentration of metals. Attention is being focused on the development of new microbial processes with the potential to recover valuable metals or to treat the wastewater. Interest also lies in selecting microbial species to provide more practical and economical methods for the potential use and reuse of microbial sources.

Microorganisms including algae, fungi and bacteria are typical examples that exhibit surface specificity towards sorption of heavy metals (Das *et al.*, 2008). Bacteria are less resistant to toxic metals than algae and fungi. Fungi and algae can be grown under extreme environmental stress conditions *i.e.*, at low pH and high temperature. Algae being autotrophic organisms require a constant supply of oxygen, photo–energy and carbon source. Moreover, due to small size, algae as well as bacteria are difficult to harvest and require laborious solid liquid separation. Fungi may be better suited as an agent for metal removal than any other microbial mass because of their great tolerance towards toxic metals and other adverse environmental conditions. They grow rapidly and often exhibit high intracellular uptake values of living cells. Moreover, fungi are microorganisms that can be economically cultivated in mass quantities under various environmental conditions. Yeasts being the unicellular fungi also have the potentiality to serve as agents for remediation of heavy metals. The major mechanisms involved in the bioremediation of heavy metals are: (a) Bioaccumulation and (b) Biosorption. Bioaccumulation is defined as the accumulation of pollutants by actively growing cells through metabolism and temperature–independent and metabolism–dependent steps (Saddetin and Donmez, 2006). Additionally, the direct utility of live cells can avoid the necessity for separate biomass production processes (*e.g.*, cultivation, harvesting, drying, processing and storage prior to use) and can be very efficient under siutable conditions that favour active growth of cells (*i.e.*, moderate pollutant concentrations, solution pH, and salt concentrations) (Wang and Hu, 2008). Biosorption is considered as promising technology for the removal of metal ions from aqueous solutions in water pollution control (Volesky, 2007). A number of studies have evaluated the application of bioaccumulation and biosorption processes for the removal of metals using yeasts. The present chapter deals with the role of yeasts as remediation agents for the removal of heavy metals from environment.

HEAVY METALS

Heavy metals are chemical elements having density above 5 g/cm^3. Of the 90 naturally occurring elements, 21 are non metals, 16 are light metals and the remaining 53 (with As included) are heavy metals which are stable elements (meaning they cannot be metabolized by the body) and bio–accumulative (passed on the food chain to humans). Heavy metals are usually classified as the following three categories: (1) toxic metals *viz.* Hg, Cr, Pb, Zn, Cu, Ni, Cd, As, Co, Sn, etc. (2) precious metals *viz.* Pt, Ag, Au, Ru etc. and (3) radionucleotides *viz.* U, Th, Ra, Am etc. (Volesky, 1990). Most heavy metals are transition elements with incompletely filled d orbitals. These d orbitals provide heavy metal cations with the ability to form complex compounds which may or may not be redox active. Thus, heavy metal cations play an important role as trace elements in sophisticated biochemical reactions (Nies, 1999). The presence of heavy metals in the environment is of major concern because of their toxicity, bio–accumulating tendency causing threat to human life and the environment (Horsfall and Spiff, 2005).

HAZARDS AND TOXICITY

Heavy metal hazard is not an entirely modern phenomenon. The health hazards caused by heavy metals depend on the level of exposure and the length of exposure. In general, exposures are divided into two classes: acute exposure and chronic exposure. Acute exposure refers to contact with a large amount of heavy metal in a short period of time, in some cases health effects are immediately apparent and in others, effects are delayed. Chronic exposure refers to contact with low levels of heavy metals over a longer period of time (Young, 2000).

Metal toxicity is the inherent capacity of the metal to affect any biological activity. The adverse effect could be an interaction of the metal with a protein or enzyme leading to changes in physiological and metabolic processes or an interaction with DNA leading to mutation and change in behavior. The toxicity is a result of an interaction between the free metal ion and the susceptible target. Among the toxic heavy metals, mercury, lead and cadmium, 'called the big three' are in the lime light due to the major impact on the environment (Volesky and Prasetyo, 1994). Hg and Cd are very toxic even in lower concentration of 0.001–0.1 mg/L (Alkorta *et al.*, 2004). Arsenic, chromium, copper and zinc are also toxic; lead and cadmium are potent neurotoxic metals (Puranik and Pakniker, 1997). The sources of human exposure to Cd include atmospheric, terrestrial and aquatic routes (Lopez *et al.*, 1994). The most severe form of Cd toxicity in humans is 'Itai–Itai', a disease characterized by excruciating pain in the bone (Yasuda *et al.*, 1995). Other health implications of Cd in human include kidney dysfunction, hepatic damage and hypertension (Klaassen, 2001). The recommended daily intake of Zn is between 4 and 16 mg depending on age, sex and physiological state (FNB, 1974). Zn is an essential element to man, being a cofactor of many enzyme systems (Ukhun *et al.*, 2005). It has been reported to competitively inhibit Pb uptake in cells (Lou *et al.*, 1991). Lead is a heavy metal poison which forms complexes with oxo–groups in enzymes to affect virtually all the steps in the process of hemoglobin synthesis and porphyrin metabolism (Ademorati, 1996). Toxic level of Pb in man has been associated with encephalopathy, seizures and mental retardation (Schumann, 1990). Copper, one of the most widely used heavy metal, mainly employed in electrical and electroplating industries is extremely toxic to living organisms. The presence of copper (II) ions, causes serious toxicological concerns. It is usually known to be deposited in brain, skin, liver, pancreas and myocardium (Davis *et al.*, 2000). The accumulation of mercury in environment, which comes from municipal wastes and manufacturing of organo–mercurial compounds, causes potential risk to human health due to the uptake of mercury by plants and their introduction into the food chain, including marine organisms (algae, seaweed, fish, etc.). The toxic and carcinogenic effects of mercury on living beings are quite well known. Mercury in liquid form is not very dangerous but in the vapour form it becomes more poisonous. It attacks the lungs, kidneys and the brain. Arsenic affects the skin causing skin cancer in its most severe form. A massive outbreak of arsenical dermatosis was reported in some parts of West Bengal which was linked with high level of arsenic in tube well waters (0.2–2.0 mg/L) (Namasivayam and Senthilkumar, 1998; Muhan and Agarwal, 2006). Arsenic occurs mainly as As (III) and As (V). Chromium has both beneficial and detrimental properties. Two stable oxidation states of chromium persists in the environment, Cr (III) and Cr (VI) which have contrasting toxicities, mobility and bioavailability (Saifuddin and Kumaran, 2005). While Cr (III) is relatively innocuous and immobile, Cr (VI) moves readily through soils and aquatic environments and is a strong oxidizing agent capable of being absorbed through the skin (Park and Jung, 2001). Hexavalent chromium, Cr (VI), is the toxic form of chromium released during many industrial processes including electro–plating, leather industry and pigment manufacture (Faisal and Hasnain, 2004). Trivalent chromium, Cr (III), is an essential element required for normal carbohydrate and lipid metabolism (Anderson, 1998). Its deficiency leads to increase in risk factors associated with diabetes and cardiovascular diseases. Contrary to deficiency symptoms, several factors make chromate contamination of intense concern, because of its toxic, carcinogenic (Shumilla *et al.*, 1999) and terartogenic (Asmatullah *et al.*, 1998) effects.

HEAVY METALS AS CARCINOGENS

A number of heavy metals are confirmed as carcinogens. Table 4.1 is a compilation of the carcinogenicity of various heavy metals.

CONVENTIONAL TECHNIQUES FOR HEAVY METAL REMOVAL

A wide range of physicochemical techniques such as chemical precipitation, reverse osmosis, ion exchange, electrodialysis, ultrafiltration and adsorption have been developed for the removal of heavy metals from wastewaters to decrease their impact on the environment.

Table 4.1: Heavy metals as human carcinogens.

Metal	*Cancers*	*Present in*	*Workers exposed*
Arsenic	Kidney, Liver, Skin, Lung	Glass, pesticides, wood	Smelting of ores containing arsenic, pesticide application, and wood preservation
Beryllium	Lung	Fibre optic products, plastic, glass, rocket fuel, nuclear weapons	Beryllium ore miners and alloy makers, nuclear reactor workers, electric and electronic, equipment workers and jewelers
Cadmium	Lung	Plastic products, batteries, fungicides	Producing, processing and handling cadmium powders, welding and re–melting of cadmium coated steel and working with solders that contain cadmium
Chromium	Lung	Automotive parts, floor covering, paper, cement, roofing, anticorrosive metal plating	Stainless steel production and welding, chromate production, chrome plating, chrome pigment and corrosive metal and tanning industries
Lead	Brain, Kidney	Cotton dyes, metal coating, drier in paints, varnishes and pigment inks	Construction work that involves welding, cutting, brazing or blasting on lead paint surfaces
Nickel	Lung, Nasal cavity	Steel, dental fillings, copper and brass, permanent magnets	Battery makers, ceramic makers, electroplaters

Chemical Precipitation

Chemical precipitation is widely used method for heavy metal removal from inorganic effluent. After pH adjustment to the basic conditions (pH 11), the dissolved metal ions are converted to the insoluble solid phase *via* a chemical reaction with a precipitant agent such as lime (Wang *et al.*, 2004). Lime or calcium hydroxide is the most commonly employed precipitant agent. Lime precipitation can be employed to effectively treat inorganic effluent with a metal concentration of higher than 1000 mg/L. Other advantages of using lime precipitation include the simplicity of the process, inexpensive equipment requirement, and convenient and safe operations, making it a popular method for metal removal from contaminated wastewater. Apart from its various advantages, it has various drawbacks like excessive sludge production that requires further treatment. The increasing cost of sludge disposal, slow metal precipitation, poor settling, the aggregation of metal precipitates, and the long–term environmental impacts of sludge disposal are other disadvantages of this process (Wingenfelder, 2005). A combined method of electro–Fenton followed by lime precipitation is used as an effective method for the treatment of rayon industry wastewater containing high COD and zinc (Ghosh *et al.*, 2011). Hydrocalumite can effectively remove Zn(II) from aqueous solutions through hydrocalumite dissolution and selective precipitation (Liu *et al.*, 2011).

Reverse Osmosis

Reverse osmosis (RO) is a pressure–driven membrane process in which water can pass through the membrane, while the heavy metal is retained. A greater hydrostatic pressure than the osmotic pressure of the feeding solution is applied to separate the cationic compounds from water (solvent). Depending on the characteristics of the membrane such as the porosity, hydrophilicity, thickness, roughness and charge of the membrane, RO works effectively at a wide pH range of 3–11 and at 4.5–15 bar of pressure (Madaeni and Mansourpanah, 2003). Pressure is the major parameter that affects the extent of heavy metal removal by RO. The metal removal efficiency increases with the increase in the level of the pressure. Various advantages of RO include a high water flux rate, high salt rejection, resistance to biological attack, mechanical strength, chemical stability and the ability to withstand high temperatures (Ujang and Anderson, 1996). The reuse of water from industrial process can be achieved and RO enables industrial users to comply with the effluent limit of the discharge standards imposed under environmental legislation. RO has some limitations such as irreversible membrane fouling and high energy consumption. Ipek (2005) investigated the removal of Zn(II) from an aqueous solution by reverse osmosis (RO) at different pH, conductivity and EDTA concentrations. It was observed that Zn(II) removal did not change much with pH, usually less than 0.88 mg/L. The addition of EDTA into the aqueous solution increased Zn(II) removal from 98.9% to 99.6% at an EDTA concentration of 240 ppm. Bakalar *et al.* (2009) used polyamide thin film composite membraneTw30–1812–50 reverse osmosis membrane to remove Cu(II), Ni(II) and Zn(II) from aqueous solution. Aljendeel (2011) studied the application of spiral wound membrane (RO membrane) for the removal of Cu(II), Ni(II), Zn(II) from industrial wastewater. This process could recover 40.8%, 41.35% and 38.44% of Cu(II), Ni(II) and Zn(II) respectively.

Ion Exchange Technology

In addition to membrane filtration, ion exchange is also one of the most frequently applied treatments worldwide for wastewater laden with heavy metals. In ion exchange, a reversible interchange of ions between the solid and liquid phases occurs, where an insoluble substance (resin) removes ions from an electrolytic solution and releases other ions of similar charge in a chemically equivalent amount without any structural change of the resin (Vigneswaran *et al.*, 2005). After separating the loaded resin, the metal is recovered in a more concentrated form by elution with suitable reagents. In general, ion exchange is effective to treat inorganic effluent with a metal concentration of less than 10 mg/L, or in the range of 10–100 mg/L, or even higher than 100 mg/L. Unlike chemical precipitation, ion exchange does not present any sludge disposal problems, thus lowering the operational costs for the disposal of the residual metal sludge. Other advantages of ion exchange include its convenience for fieldwork since the required equipment is portable, the speciation results are reliable and the experiments can be done quickly. Resins also have certain ligands that can selectively bind with certain metal cations, making ion exchange easy to use and less time–consuming (Korngold, 2003). Despite these advantages, ion exchange also has some limitations in treating wastewater laden with heavy metals. Prior to ion exchange, appropriate pretreatment systems for secondary effluent such as the removal of suspended solids from wastewater are required. In addition, suitable ion exchanger resins are not available for all heavy metals and the capital and operational cost is high. A number of investigators reported that the removal of zinc ions using ion exchange resin showed good potential for industrial wastewater treatment (Jha *et al.*, 2008). Rotating cylindrical ion exchange bed reactor packed with cationic exchange resin is an efficient tool for removing Zn(II) from wastewater (Abdelwahab *et al.*, 2013).

Electrocoagulation

Electro–coagulation may be a better alternative than the conventional coagulation, as it can remove the smallest colloidal particles and produce just a small amount of sludge (Elimelech and Melia, 1990). However, this technique also creates a floc of metallic hydroxides which requires further purification making the recovery of valuable heavy metals impossible. The performance of electrocoagulation with aluminium elelectrodes was tested for simultaneous removal of Ni(II), Cu(II), Zn(II) and Cr(VI) from synthetic solution and electroplating wastewater (Dermentzis *et al.*, 2011). The reduction of heavy metal concentration below the permissible limit in 60 mins was noted which proved the process as safe, reliable and cost effective for heavy metal removal.

Ultrafiltration

Ultrafiltration (UF) utilizes permeable membrane to separate heavy metals, macromolecules and suspended solids from inorganic solution on the basis of the pore size (5–20 nm) and molecular weight of the separating compounds (1000–100,000 Da) (Vigneswaran *et al.*, 2005). These unique specialties enable ultrafiltration to allow the passage of water and low–molecular weight solutes, while retaining the macromolecules having a size larger than the pore size of the membrane. Depending on the membrane characteristics, UF can achieve more than 90% removal efficiency with a metal concentration ranging from 10 to 112 mg/L at pH ranging from 5 to 9.5 and at 2–5 bar of pressure. UF presents some advantages such as lower driving force and a smaller space requirement due to its high packing density. However, the decrease in UF performance due to membrane fouling has hindered it from a wider application in wastewater treatment. Fouling has many adverse effects on the membrane system such as flux decline, an increase in transmembrane pressure (TMP) and the biodegradation of the membrane materials (Choi *et al.*, 2005). These effects result in high operational costs for the membrane system. Aguirre *et al.* (2010) reported the application of micellar enhanced ultrafiltration (MEUF) combined with SDS surfactant for the removal of Cd(II) and Zn(II) from their respective water samples. They improved and optimized the process by response surface methodology (RSM). Some interesting findings were reported by Monem *et al.* (2011) who used rhamnolipid biosurfactants in MEUF for removal of Cu(II), Zn(II), Ni(II), Pb(II) and Cd(II) from metal refining wastewater.

Nanofiltration

Nanofiltration (NF) is a form of filtration that uses membranes to separate different fluids or ions. Due to its specified membrane pore structure as compared to the membranes used in RO, NF allows more salt passage through the membrane. Because it can operate at much lower pressures, and passes some of the inorganic salts, NF is used in applications where high organic removal and moderate inorganic removals are desired. NF is capable of concentrating sugars, divalent salts, bacteria, proteins, dyes and other constituents that have a molecular weight greater than 1000 daltons.

Microfiltration

This is by far the most widely used membrane process with total sales greater than the combined sales of all other membrane processes. Microfiltration (MF) has numerous small applications. It is essentially a sterile filtration with pores (0.1–10.0 microns), so small microorganisms can not pass through them. The MF membranes are made from natural or synthetic polymers such as cellulose nitrate or acetate, polyvinylidene difluoride (PVDF), polyamides, polysulfone, polycarbonate, polypropylene etc. MF has a wide array of applications such as chemical industry, food and beverages, microelectronic industry, fermentation, laboratory analytical uses etc.

Electrodialysis

This is an electro–membrane process in which ions are transported through a membrane from one solution to another under the influence of an electrical potential. Electrodialysis (ED) is widely used for production of potable water from sea or blackish water, electroplating rinse recovery, desalting of cheese whey, production of ultrapure water etc. ED membranes are usually made of cross linked polysterene that has been sulfonated. The system consists of two kinds of membranes: cation and anion, which are placed in an electric field. The cation–selective membrane permits only the cations, and anion selective membranes permit only the anions.

BIOSORPTION: AN ALTERNATIVE TECHNOLOGY FOR HEAVY METAL REMOVAL

Conventional processes used for effluent treatment are mostly expensive and not eco–friendly. The main disadvantages of the conventional treatment techniques are the generation of sludge, incomplete and long term removal of metals. The search for new cost effective technology for removal of heavy metals from wastewaters has directed attention to biosorption which is known for a last few decades. The broad definition of the term 'biosorption' is a process whereby microorganisms, alive or dead, or their derivatives are employed for the sequestration of heavy metals from environment. Biosorption has emerged as an alternative and sustainable strategy as it is effective, cheap and eco–friendly for cleaning up water that has been contaminated with heavy metals by anthropogenic activities and/or by natural processes (Suazo–Madrid *et al.*, 2011). Algae, bacteria, fungi and yeasts have been reported as potential metal biosorbents which possess metal sequestering properties and can decrease the concentration of heavy metals. Compared to conventional methods, biosorption offers several advantages. The major advantages of biosorption include low cost, high efficiency, minimization of chemical and biological sludge, no additional nutrient requirement, regeneration of biosorbent and possibility of metal recovery (Chatterjee *et al.*, 2010; Zhang *et al.*, 2010).

YEAST AS REMEDIATION AGENT

Metal Removal Through Biosorption

Among the promising agents for heavy metal removal which have been searched during the past decades, yeast (*Saccharomyces cerevisiae)* has received increasing attention due to the unique nature in spite of its mediocre capacity for metal uptake. In *S. cerevisiae*, free cells appeared unsuitable in practical application, largely due to solid/liquid separation problem. However, Veglio and Beolchini (1997) pointed out that investigation on the performance of free cells for metal uptake can provide fundamental information on the equilibrium of the biosorption process, which is useful for practical application. Meanwhile, flocculating cell has been suggested for biosorption, attempting to overcome the separation problem of free cells (Soares *et al.*, 2002). The yeast cells of *S. cerevisiae* treated with hot alkali were capable of accumulating a wide range of heavy metal cations (Fe^{3+}, Cu^{2+}, Cr^{3+}, Hg^{2+}, Pb^{2+}, Cd^{2+}, Co^{2+}, Ni^{2+} and Fe^{2+}. Some toxic metal ions studied in *S. cerevisiae* biosorption are lead (Goksungur *et al.*, 2005; Ozer and Ozer, 2003), copper (Bakkaloglu *et al.*, 1998), zinc (Bakkaloglu *et al.*, 1998), cadmium (Goksungur *et al.*, 2005, Gomes *et al.*, 2002; Park *et al.*, 2003; Vasudevan *et al.*, 2003), mercury (Zhu *et al.*, 2004), cobalt (Al Saraj *et al.*, 1999), nickel (Bakkaloglu *et al.*, 1998; Ozer and Ozer, 2003) and chromium (Nguyen–nhu and Knoops, 2002). Biosorption of chromium (VI) and arsenic (V) onto methylated yeast biomass has been studied by Seki *et al.* (2005). *S. cerevisiae* can distinguish different metal species such as selenium (IV) and selenium (VI), antimony (III) and mercury (II) based on their toxicity. This kind of property makes *S. cerevisiae* useful not only for the removal and recovery of metal ions, but also for their analytical measurement (Perez–Corona *et al.*, 1997). Equilibrium and thermodynamic studies on biosorption of Pb (II) onto *Candida albicans* biomass was conducted by Baysal *et al.* (2009). Machado *et al.* (2010) examined the influence of the competitive effect of inorganic ligands (carbonates, chlorides, uorides, phosphates, nitrates and sulphates) on biosorption of Zn(II) ions alongwith other metals by yeast cells of *S. cerevisiae. Saccharomyces cerevisiae* was successfully used for biosorption of some heavy metal ions from aqueous and nonaqueous samples. This behavior was mainly due to the strong capability of yeast cells to bind with the interacting metal ions Ca^{2+}, Cd^{2+} and Pb^{2} *via* their metal chelating and ion exchange characters by using the loaded functional groups. Ahmad *et al.* (2013) focused on the use of immobilized biomass of *Candida utilis* and *Candida tropicalis* for the enhanced biosorption of zinc from aqueous solution. Das *et al.* (2012) reported maximum Zn(II) removal by two yeast isolates *Candida* sp. VITGBN1 and *Cryptococcus* sp. VITGBN2 at an optimized pH 6.0; biosorbent dosage 1.5 g/L; initial Zn(II) concentration 90 mg/L and contact time 240 min. Column studies were conducted by alginate immobilized *Candida* sp. VITGBN1 packed in column. Maximum Zn(II) removal was noted at a bed height of 12 cm, flow rate 1 mL/min and initial Zn(II) concentration of 150 mg/L. Column adsorption data were applied to BDST model. Finally attempts were made to regenerate and reuse the immobilized SDS treated dead yeast using column mode of operation. Three cycles of desorption and sorption was performed to study the reusability of the biosorbent. The elution efficiency and Zn(II) removal were found to decrease in the successive cycles. This behavior was primarily due to continuous usage of the biosorbent.

Basak and Das (2013) reported that pretreatment of yeast biomass with anionic surfactant enhanced the removal of Zn(II) ion to a greatest extent compared to other treated and untreated biosorbents. In this study, treatment of yeast biomass with anionic surfactants reduced its total acidity and increased its total basicity. This result implied that these anionic surfactants successfully covered the surface of the yeast biomass. Total acidity of the yeast biomass was decreased because the surfactants covered surface acidic groups, especially carboxylic groups. The hydrophilic head(s) of the surfactants can act as basic functional groups. In aqueous solutions, the bound anionic surfactants can be dissociated, and then the protons bind to the hydrophilic head(s), resulting an increase of the adsobents total basicity (Ahn *et al.*, 2009). Among the various organic solvents used for zinc(II) removal, gluteraldehyde treated yeast biomass showed improvement in Zn(II) removal followed by methanol and formaldehyde. The treatment of biomass with methanol caused esterification of the carboxylic acid present on the cell wall. The metal binding ability of carboxyl groups was reduced as a result of esterification (Drake *et al.*, 1996). In case of formaldehyde treatment, the results revealed that amino groups present on the cell wall of yeast biomass gets methylated due to chemical modification with formaldehyde. Thus the methylation of amino groups reduced the binding of metal ions on the biomass residue (Loudon, 1984).

Quantitative study and structural characterization of uranium uptake by both live and heat–killed *Saccharomyces cerevisiae* at environmentally relevant uranium concentration and with different ionic strengths were carried out by Lu *et al.* (2013). Kinetic investigation showed that the equilibrium reached within 15 min. In equilibrium studies, pH shift towards neutral indicated release of hydroxyl ions. pH was the most important factor, which partly affected electrostatic interaction between uranyl ions and *S. cerevisiae* surface. The high ionic strength inhibited biosorption capacity, which could be explained by a competitive reaction between sodium ions and uranyl ions. Heat killing process significantly enhanced the biosorption capacity, showing an order of magnitude higher than that of live cells. High resolution transmission electron microscopy (HRTEM) coupled with energy dispersive X–ray (EDX) showed needle–like uranium–phosphate precipitation formed on the cell walls for both live and heat–killed cells. Besides, dark–field micrographs displayed considerable similar uranium–phosphate precipitation presented outside the heat–killed cells. The phosphate was released during heat–killing process. FTIR illustrated the involvement of hydroxyl, carboxyl, phosphate, and amino functional groups which played an important role in complexation with uranium.

Enhanced bisorption of zinc ions from aqueous solution by immobilized *Candida utilis* and *Candida tropicalis* cells was reported

(Ahmed *et al.*, 2013). The isotherms, kinetics and thermodynamics of biosorption were studied in batch system under optimum operating conditions. Experimental equilibrium data were analyzed using Langmuir, Freundlich and Dubinine Raduskhkevich isotherm models by linear and non–linear regression. The functional groups responsible for the biosorption of zinc ions were identified using Fourier transform infrared spectra. Immobilized *C. utilis* exhibited the highest biosorption capacity of 181.7 mg/g at 45°C for removal of zinc ions from aqueous solution. The kinetic analysis of the experimental data showed that the biosorption process followed the pseudo second order kinetics. Different desorbing agents were investigated for elution of adsorbed zinc ions from the immobilized biomass and regenerated biosorbent was reused without significant loss in its biosorption capacity. Thermodynamic parameters were calculated using the biosorption equilibrium constant based on Langmuir isotherm.

Biosorption of Cr(VI) from aqueous and real water samples was reported using gelatin–impregnated–yeast as a novel biosorbent (Mahmoud, 2015). Gelatin and yeast materials (GeleYst) were combined together to form a novel eco–friendly, non–toxic, non–carcinogenetic, biodegradable, biocompatible and inexpensive biosorbent to enhance the extraction and biosorption of Cr(VI) from water. The potential of GeleYst for biosorption of Cr(VI) was studied by the static technique in various buffer solutions. Other controlling experimental factors were also examined and evaluated. The applications of GeleYst biosorbent in water treatment processes of Cr(VI) from real acidic and neutral water samples were successfully accomplished using multi–stage micro–column techniques.

Metal Removal Through Bioaccumulation

Bioaccumulation is defined as the accumulation of pollutants by actively growing cells by metabolism and temperature–independent and metabolism–dependent mechanism steps (Saddetin and Donmez, 2006). Accumulation of zinc by wine yeast strains of *Saccharomyces cerevisiae* during fermentation of both grape juice and chemically defined medium with different carbohydrates and at varying levels of zinc was studied (Nicola *et al.*, 2009) The results showed that zinc accumulation by wine yeast was very rapid with all zinc being removed from the medium by yeast cells within the first two hours. Zinc uptake was stimulated by the presence of sucrose. Zinc affected fermentation progress at defined levels, with optimal concentrations at 1.5–2.5 ppm, depending on yeast strain and zinc bioavailability. Five yeast species was evaluated for the bioaccumulation of Zn(II) ions (Roepcke *et al.*, 2011). The highest Zn concentration was 6820 mg/kg of dry weight biomass using *Pichia guilliermondii* Wickerham LPB 063 after 120 h of cultivation in a medium with 0.5 g/L $ZnSO_4$. Basak *et al.* (2011) reported that two yeast species *viz Candida* sp. VITGBN1 and *Cryptococcus* sp. VITGBN2 isolated from industrial wastewater were capable of accumulating Zn(II) ion using extract of sugarcane bagasse which served as the sole source of nutrient in the medium. Use of aqueous extract of sugarcane bagasse as a cost effective growth medium was one of the attractive features of the study. *Candida* sp. VITGBN1 was found to be more efficient in bioaccumulating Zn(II) ion than *Cryptococcus* sp. VITGBN2 at all concentrations of Zn(II) ions. Bioaccumulation of Zn(II) ion was increased with increasing concentration of sugar extracted from sugarcane bagasse which reduced the inhibition effect of Zn(II) ion on yeasts growth. Inhibition kinetics showed that *Candida sp.* VITGBN1 is more resistant to Zn(II) ion than *Cryptococcus* sp. VITGBN2. The study also revealed that the bioaccumulation of Zn(II) ion using yeasts followed first–order–kinetics depicting that the process was dependent on the concentration of Zn(II) ions in solution. A slight variation in the concentration resulted change in Zn(II) uptake by yeast cells.

Application of *Zygosaccharomyces rouxii* and *Saccharomyces cerevisiae* for the bioaccumulation of Zn(II) ions in presence of high salt (NaCl) concentration was reported (Li *et al.*, 2013). The Zn(II) bioaccumulation capacities of *Z. rouxii* and *S. cerevisiae* increased after the addition of NaCl which suggested that NaCl enhanced the Zn(II) uptake of salt tolerant yeasts. For both the yeasts, the amount of Zn(II) was removed intracellularly. The rate of zinc removal increased from

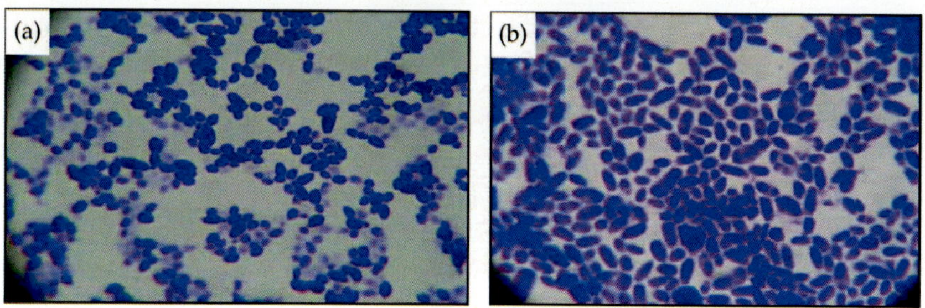

Fig. 4.1: Microscopic images of (a) *Candida* sp. VITGBN1 (b) *Cryptococcus* sp. VITGBN2

94.4% to 96.5% and 50.2% to 50.7% by *Z. rouxii* and *S. cerevisiae* respectively at initial Zn(II) concentration of 0.1 mmol/L after the addition of 10 g/L NaCl.

Metal Removal Using Biosurfactant Produced by Yeast

The production of biosurfactants (lactonic monoacetate and acidic diacetate sophorolipid) by *Candida* sp. VITGBN1 and *Cryptococcus* sp. VITGBN2 and their application for Zn(II) removal from electroplating wastewater was reported. Maximum biosurfactants production was obtained at pH 7.0, temperature of 30°C and substrate concentration of 2% (v/v) vegetable oil substrate. At the initial stationary phase, higher concentration of biosurfactants was obtained. The biosurfactant production was screened based on cell surface hydrophobicity, drop collapse test, oil displacement, surface tension and emulsification index. FT–IR spectra and MS spectra confirmed the involvement of functional groups and chemical structure confirmed the lactonic monoacetate sophorolipid produced by *Candida* sp. VITGBN1 and acidic diacetate sophorolipid produced by *Cryptococcus* sp. VITGBN2 (Basak and Das, 2014). Maximum Zn(II) removal was noted at 100 mg/L of Zn(II) ion concentration. At a concentration higher than 100 mg/L, Zn(II) removal was decreased. This might be explained due to the formation of co–precipitate of biosurfactant and Zn(II) ions together. Figure 4.2 shows the structure of the biosurfactants produced by *Candida* sp. VITGBN1 and *Cryptococcus* sp. VITGBN2 respectively.

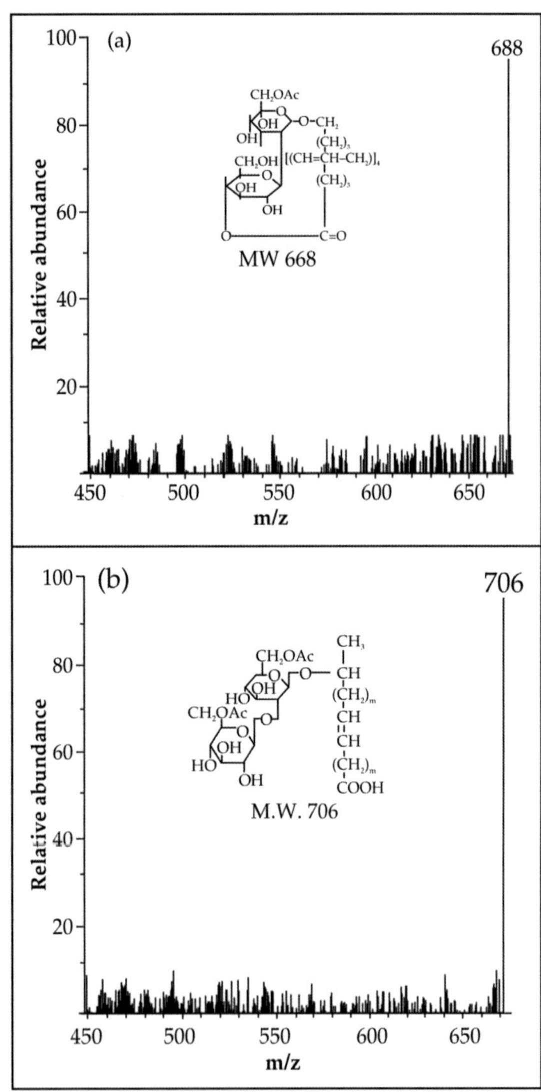

Fig. 4.2: Structure of the biosurfactants produced by *Candida* sp. VITGBN1 and *Cryptococcus* sp. VITGBN2.

The results of the FT–IR spectral analysis of the native biosurfactants are shown in Fig. 4.3a and Fig. 4.4a. FT–IR spectra of biosurfactant interacted with Zn(II) ions showed that the peaks expected at 3435.34,

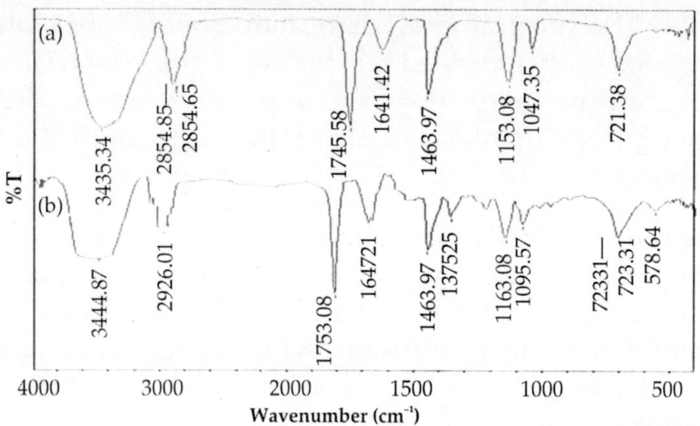

Fig. 4.3: FTIR spectra of *Candida* sp. VITGBN1 (a) native biosurfactant and (b) Zn(II) interacted biosurfactant.

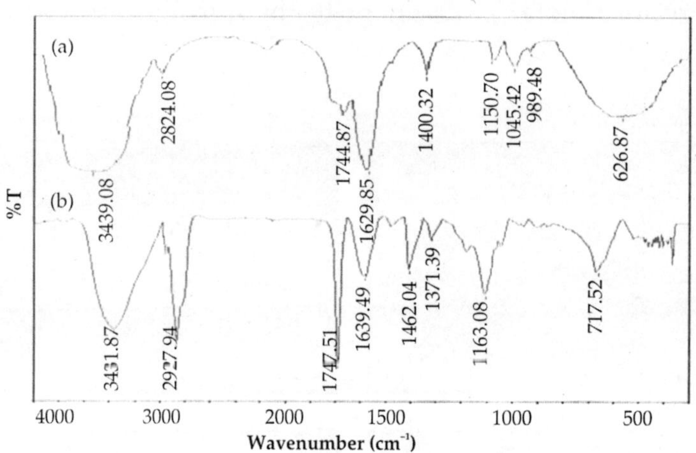

Fig. 4.4: FTIR spectra of *Cryptococcus sp.* VITGBN2 (a) native biosurfactant and (b) Zn(II) interacted biosurfactant.

1745.58, 1641.42 and 1047.35 cm^{-1} had shifted to 3444.87, 1763.08, 1647.21 and 1095.57 cm^{-1} for *Candida* sp. VITGBN1 biosurfactant (Fig. 4.3b) and the peaks expected at 3439.08, 1744.87, 1629.85 and 1400.32 cm^{-1} were shifted to 3437.81, 2927.94, 1747.51, 1639.49, 1426.04 and 1163.08 cm^{-1} for *Cryptococcus* sp. VITGBN2 biosurfactant

(Fig. 4.4b). The spectral analysis of native and Zn(II) interacted biosurfactant clearly indicated that hydroxyl group (–OH), carboxyl (–COOH) and carbonyl (C=O) groups were the predominant functional groups involved in zinc removal by both the type of biosurfactants.

The removal of Zn(II) ion from the aqueous solution was further confirmed using SEM analysis. Fig. 4.5(a) showed the morphology of native biosurfactants, showing the self–assembly of sophorolipid molecule conforming to hexagonal structure for both the yeast isolates. Figure 4.5(b) showed sequestered Zn(II) ions onto biosurfactants. The projected spherical nodules in the SEM images confirmed the anchoring of Zn(II) ions with the biosurfactant molecule. EDS of the native biosurfactants (Fig.4.6a) and Zn(II) ion interacted biosurfactants (Fig. 4.6b) served as a direct proof of metal attachment to the micellar structure of the biosurfactants.

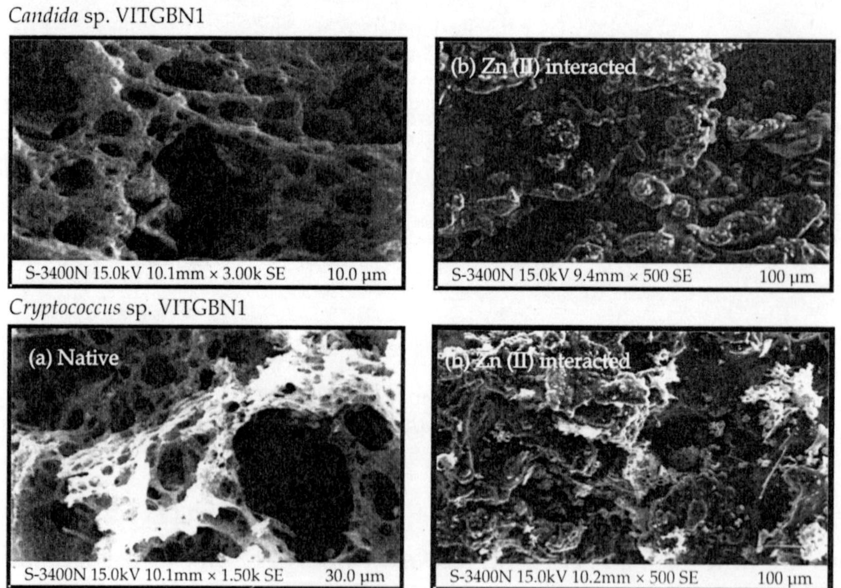

Fig. 4.5: SEM images of (a) native biosurfactant (b) Zn(II) ions interacted biosurfactants (Basak and Das, 2014).

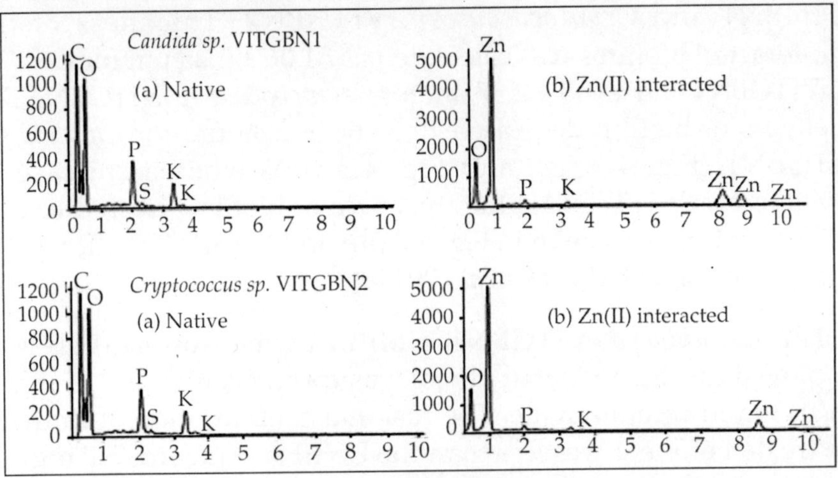

Fig. 4.6: EDS analysis of (a) native biosurfactant and (b) Zn(II) interacted biosurfactants (Basak and Das, 2014).

METAL REMOVAL USING YEAST BIOFILM

The biofilm forming capability of *Candida sp.* VITGBN1 and *Cryptococcus sp.* VITGBN2 on gravels over 72 h were reported by Basak *et al.* (2014a). The production of the soluble coloured formazan salt from biofilm forming yeast cells on gravels was a direct proof of cellular metabolic activity which increased with time. The sequence of biofilm formation by the two yeast isolates in different time intervals was clearly seen in CLSM images. Thus it was evident that yeast species were capable of forming single–species biofilm. CLSM imaging of uninteracted (i) 2D image (ii) 3D image and Zn(II) interacted (iii) 2D image (iv) 3D image of *Candida sp.* VITGBN1 biofilm and *Cryptococcus sp.VITGBN2* after 72 h of growth is shown below (Fig. 4.7 and Fig. 4.8) Formation of yeast biofilms were analysed by CLSM. The uninteracted biofilms on PVC strip showed more thickness of biomass than the thickness of Zn(II) interacted biofilms. Fig. 4.7 (i) and Fig. 4.8 (i) showed average projections of CLSM z–series images from uninteracted biofilms formed by *Candida sp.* VITGBN1 and *Cryptococcus sp.* VITGBN2 respectively. Fig. 4.7 (ii) and Fig. 4.8 (ii) showed the thickness of biofilm with z–stacks three–dimensional (3D) reconstruction respectively for *Candida sp.*

VITGBN1 and *Cryptococcus sp*. VITGBN2. Thickness of the uninteracted biofilms was calculated as 51.06 ± 1.25 μm for *Candida sp*. VITGBN1 and 45.03 ± 2.37 μm for *Cryptococcus sp*. VITGBN2 The thickness of biofilm decreased to 37.60 ± 2.89 μm for *Candida sp*. VITGBN1 (Fig. 4.7 (iii) and Fig. 4.8 (iv)) whereas in case of *Cryptococcus sp*. VITGBN2 the thickness of biofilm was reduced to 34.78 ± 3.54 μm as shown in Fig. 4.7 (iii) and Fig. 4.8 (iv) after Zn(II) interaction for 12 h (Basak *et al*., 2014a).

The *Candida sp*. VITGBN1 biofilm formed on gravels was employed for the treatment of real wastewater containing 85 mg/L Zn(II) ion in column mode. The residual concentration of Zn(II) in the treated effluent was less than the US EPA standard (5.0 mg/L). The physicochemical characteristics of the effluent before and after treatment in the column are shown in the Table 4.2. There was significant difference in the physicochemical properties of industrial wastewater after treated with yeast biofilm. The residual concentrations of other metals *viz*. Ni(II), Cd(II) and Cu(II) were found to be 2.8 mg/L, 1.8 mg/L and 2.6 mg/L respectively. These residual concentrations were less than the permissible limits of each metal.

Table 4.2: Physico–chemical analysis of electroplating wastewater treated with *Candida sp*.VITGBN1 biofilm in packed bed column (Basak *et al*., 2014a).

Parameter	*Before treatment*	*After treatment*
pH	7.63 ± 0.17	6.4 ± 0.63
Conductivity (μΩ)	5.68 ± 0.20	2.1 ± 0.02
TSS (mg/L)	1310 ± 3.7	615 ± 5.2
TDS (mg/L)	1137 ± 4.1	550 ± 4.5
COD (mg/L)	61 ± 1.2	23 ± 0.64
Zn(II) (mg/L)	85 ± 0.64	4 ± 0.4
Cd(II) (mg/L)	10 ± 0.54	1.8 ± 0.09
Ni(II) (mg/L)	21 ± 0.76	2.8 ± 0.17
Cu(II) (mg/L)	20 ± 0.43	2.6 ± 0.15

Dual role of sophorolipid produced by the yeast *Cryptococcus sp*.VITGBN2 as biostabilizer for ZnO nanoparticle synthesis and biofunctionalizing agent against *Salmonella enterica* and *Candida*

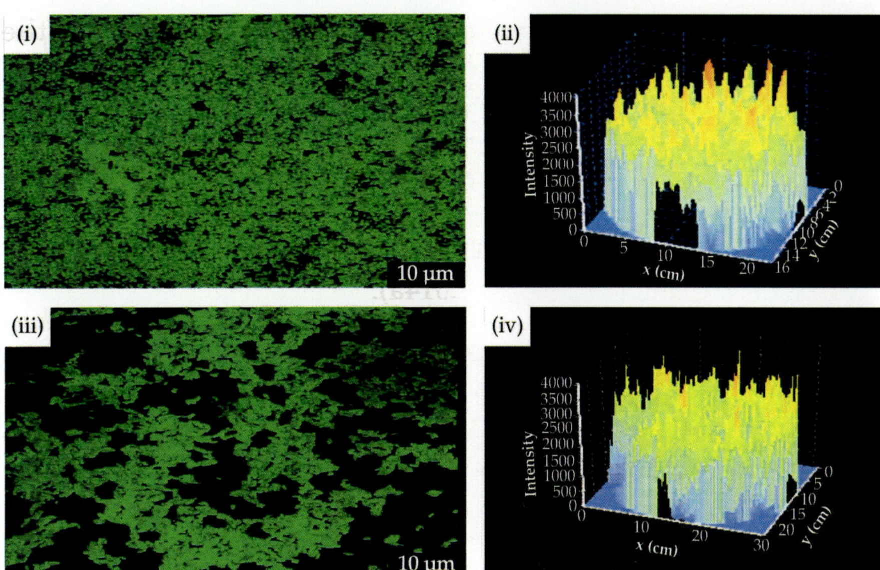

Fig. 4.7: CLSM imaging of uninteracted (i) 2D image (ii) 3D image and Zn(II) interacted (iii) 2D image (iv) 3D image of *Candida* sp. VITGBN1 biofilm after 72 h of growth (Basak *et al.*, 2014a).

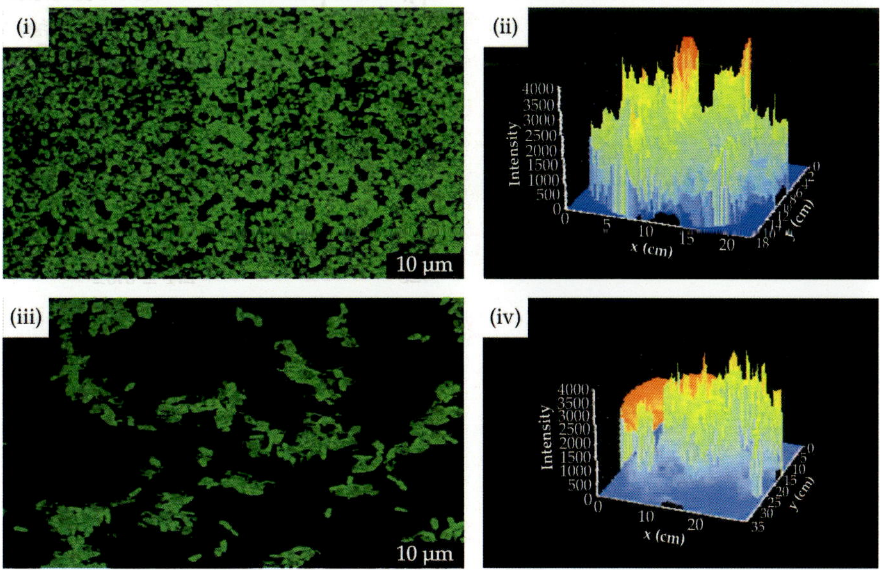

Fig 4.8: CLSM imaging of uninteracted (i) 2D image (ii) 3D image and Zn(II) interacted (iii) 2D image (iv) 3D image of *Cryptococcus* sp. VITGBN2 biofilm after 72 h of growth (Basak *et al.*, 2014 a).

albicans has been reported by Basak *et al.* (2014b). The formation of biofunctionalized ZnO nanoparticle was characterized using UV spectroscopy, XRD, Scanning electron microscopy (SEM) and Fourier Transform Infrared spectroscopy (FTIR). The antimicrobial activities of naked ZnO nanoparticle and biofunctionalized ZnO nanoparticle were tested based on the diameter of inhibition zone in agar well diffusion assay. Microbial growth rate determination, protein leakage diffusion assay, and lactate dehydrogenase assay was done. Bacterial pathogen *Salmonella enteric* and fungal pathogen *Candida albicans* showed more sensitivity to sophorolipid biofunctionalized ZnO nanoparticle compared to naked ZnO nanoparticle. This was the first report on dual role of yeast mediated sophorolipid used as a stabilizer for ZnO nanoparticle synthesis as well as a novel biofunctionalizing agent showing antimicrobial property.

5

Role of Yeasts for Remediation of Synthetic Dyes

INTRODUCTION

Synthetic dyes are used widely in industries such as paper, colored photography and textile. There are 10,000 types of dyestuff throughout the world and approximately 7×10^5 tons of these are produced every year. Generally, synthetic dyes can be classified as anionic (direct, acid and reactive dyes), cationic (basic) and non–ionic (disperse). Anionic dyes represent 20–30% of commercial dyes used. Reactive dyes are formed by the combination of azo–based chromophores with different types of reactive groups such as vinyl sulfone, chlorotriazine, trichloropyrimidine and difluorochloropy–rimidine. Azo dyes are the largest chemical class of dyes with a great deal of structural and color variety used in industries representing up to 70% of the annual production. They are characterized by the presence of one or more azo groups which are responsible for their coloration, recalcitrant nature and hence are less biodegradable. Reactive dyes are commonly used in textile industries because of their favorable characteristics such as bright color water fastness, and their simple application techniques with low energy consumption (Tan *et al.*, 2013).

There are various physical and chemical wastewater treatment methods that can be applied for textile wastewater treatment such as flocculation, coagulation, adsorption, membrane filtration, precipitation, irradiation and ozonation (Banat *et al.*, 1996; Slokar and Le Marechal, 1998). But the applicability of those methods is limited due to various limitations (Wijetunga *et al.*, 2007). Advanced oxidation process such as Fenton's reagent (H_2O_2 and Fe^{2+}), H_2O_2, and Ozonation are costly in terms of operational costs (Stanislaw and Monika, 1999). Coagulation and flocculation with lime, alum, polyelectrolyte, and ferrous salts produce large amount of sludge, which impose handling and disposal problems (Huseyin, 2005). The adsorption of dyes using activated carbon and various other adsorbents are also costly (Zhemin *et al.*, 2001). The photochemical oxidation of dyes using UV and sunlight with oxidation agents like catalyses, H_2O_2, is also not economically viable (Muruganandham and Swaminathan, 2004).

Microorganisms can accumulate high concentration of dyes. Attention is being focused on the development of new microbial processes with the potential to remove dyes for the treatment of wastewater (Parshetti *et al.*, 2007). The different mechanisms involved in the bioremediation of textile dyes are: (a) *Bioaccumulation*: in intracellular milieu and their further assimilation in living beings specifically microbes (Kuhn and Pfister, 1990), (b) *Biodegradation*: by direct and indirect intracellular or extracellular enzymes (Blanquez *et al.*, 2004; Dawkar *et al.*, 2010) and (c) *Bioadsorption*: through ion exchange by living and dead microbial biomass (Zhou and Zimmerman, 1993).

Over the past decade, many fungal and bacterial species have been reported for their abilities to remediate a variety of structurally diverse synthetic dyes. One of the most ubiquitous biomass types available for bioremediation of textile dyes is yeast. Yeasts are inexpensive, readily available sources of biomass. Furthermore, yeast cells retain their ability to accumulate a broad range of textile dyes to varying degrees under a wide range of external conditions (Donmez, 2002; Aksu, 2003). The potentiality of yeast species as remediation agent for the removal of synthetic dyes have been discussed in this chapter.

CLASSIFICATION OF SYNTHETIC DYES

Synthetic dyes became popular because they are easy–to–use, less expensive and have wider range of colors. Synthetic dyes are obtained by adding chemicals to natural dyes. For example, artificial alizarin is a synthetic dye synthesized from coal tar anthracene. Synthetic dyes are used for modern clothing.

Classification Based on Dying Process

Dyeing process is accomplished by dissolving or dispersing the colorant in a suitable solution (usually water) and bringing this system into contact with the material to be dyed. According to the process of dying, dyes are classified into the following types:

Acid dyes

Acid dyes are water–soluble anionic dyes that are applied to the materials such as silk, wool, nylon and modified acrylic fibres using neutral to acid dye baths. Attachment to the fibre is attributed to salt formation between anionic groups in the dyes and cationic groups in the fibre. Acid dyes are not substantive to cellulosic fibres. Most synthetic food colors fall in this category.

Basic dyes

Basic dyes are water–soluble cationic dyes that are mainly applied to acrylic fibres, but find some use for wool and silk. Usually acetic acid is added to the dye bath to help the uptake of the dye onto the fibre. Basic dyes are also used in the coloration of paper.

Direct dyes

Direct dying is normally carried out in a neutral or slightly alkaline dyebath, at or near boiling point, with the addition of either sodium chloride (NaCl) or sodium sulfate (Na_2SO_4). Direct dyes are used on cotton, paper, leather, wool, silk and nylon. They are also used as pH indicators and as biological stains.

Mordant dyes

These dyes require a mordant, which improves the fastness of the dye against water, light and perspiration. The choice of mordant is very important as different mordants can change the final color significantly. Most natural dyes are mordant dyes and there is a large literature base describing dyeing techniques. The most important mordant dyes are the synthetic mordant dyes or chrome dyes, used for wool. These comprise some 30% of dyes used for wool, and are especially useful for black and navy shades. The mordant, potassium dichromate, is applied after–treatment. It is important to note that many mordants, particularly those in the heavy metal category, can be hazardous to health and extreme care must be taken in using them.

Vat dyes

These dyes are essentially insoluble in water and incapable of dyeing fibres directly. However, reduction in alkaline liquor produces the water soluble alkali metal salt of the dye, which has an affinity for the textile fibre. Subsequent oxidation reforms the original insoluble dye. The color of denim is due to indigo, the original vat dye.

Reactive dyes

These dyes utilize a chromophore attached to a substituent that is capable of directly reacting with the fibre substrate. The covalent bonds that attach reactive dye to natural fibres make them among the most permanent of dyes. "Cold" reactive dyes, such as Procion MX, Cibacron F, and Drimarene K, are very easy to use because these dyes can be applied at room temperature. Reactive dyes are by far the best choice for dyeing cotton and other cellulose fibres at home or in the art studio.

Disperse dyes

Disperse dyes were originally developed for the dyeing of cellulose acetate, and are water insoluble. The dyes are finely ground in the presence of a dispersing agent and sold as a paste, or spray–dried and sold as a powder. Their main use is to dye polyester but they can also be used to dye nylon, cellulose triacetate, and acrylic fibres.

In some cases, a dyeing temperature of 130°C is required, and a pressurised dyebath is used. The very fine particle size gives a large surface area that aids dissolution to allow uptake by the fibre. The dyeing rate can be significantly influenced by the choice of dispersing agent used during the grinding.

Sulfur dyes

These are two part "developed" dyes used to dye cotton with dark colors. The initial bath imparts a yellow color and is treated with a sulfur compound in place to produce the dark black color. Sulfur Black 1 is the largest selling dye by volume.

Classification Based on Chemical Structure

Acridine dyes

Acridine dyes are derived from acridine (Fig. 5.1). They have the following structure:

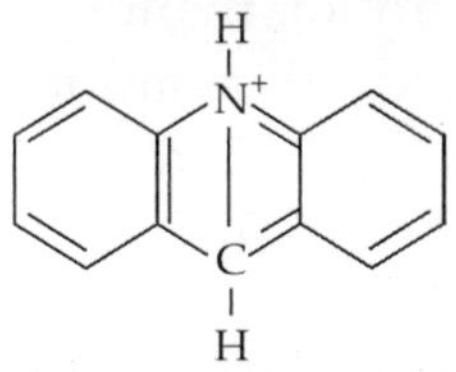

Fig. 5.1: General structure of acridine dyes.

The general formula shows the chromophore to be a quinoid ring. The molecule is positively charged (basic) as the nitrogen atom is attached to the quinoid ring. The commonest example would be acridine orange.

Anthraquinone dyes

Anthraquinone dyes (Fig. 5.2) are derived from anthracene. They have the general formula. The general formula shows the chromophore to be a quinoid ring. This class of dyes may have either hydroxyl groups or amino groups attached to the general structure. Alizarin and alizarin red are examples of anthraquinone dyes.

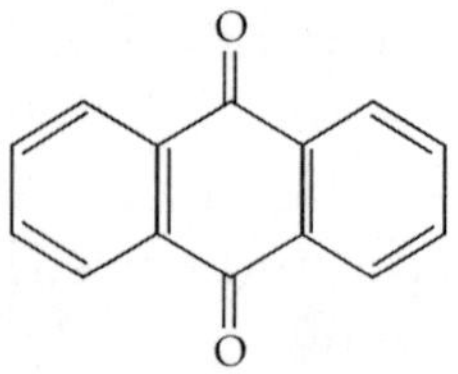

Fig. 5.2: General structure of anthraquinone dyes.

Arylmethane dyes

Arylmethane dyes are so called because they are derived from methane in which some of the hydrogen atoms are replaced with aryl rings. Since a synonym for aryl is phenyl, they may also legitimately be called phenylmethane dyes. Aryl rings are often referred as benzene rings. However, benzene is a specific chemical compound and this usage is technically incorrect, although very common.

Diarylmethanes

Diarylmethane (or diphenylmethane) dyes have two aryl rings. Auramine O, a fluorescent dye is example of diarylmethane dye. Its structural formula is shown below. The alkylamino groups are attached to the aryl rings (Fig. 5.3).

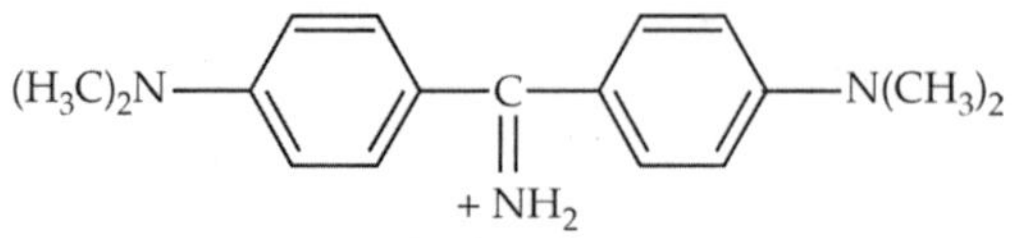

Fig. 5.3: Chemical structure of auramine O.

Triarylmethanes

Triarylmethane (or triphenylmethane) dyes contain three aryl rings. The pararosaniline homologues are good examples of triarylmethane dyes. These dyes also contain amino groups, and are identified as aminotriarylmethane dyes, and are often basic. In pararosanilin (Fig. 5.4), each of the three aryl rings has an amino group. The quinoid ring is the chromophore.

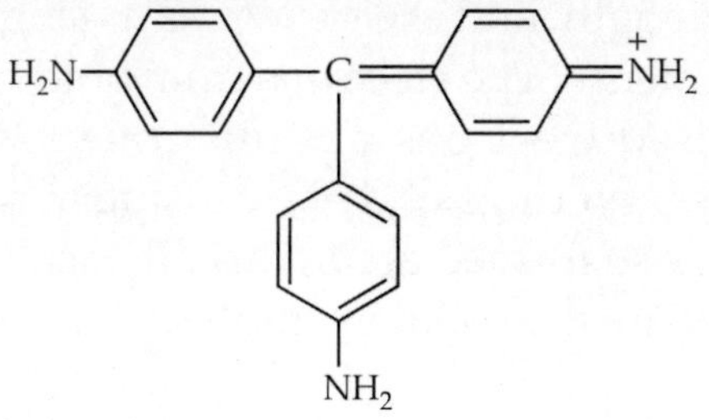

Fig. 5.4: Chemical structure of pararosaniline homologues.

Azo dyes

Azo dyes contain the azo chromophore (–N=N–). They have the following general structure (Fig. 5.5). They differ in auxochromes. Common auxochromes are $–NH_2$, NR_2 and –OH groups. Examples are aniline yellow, butter yellow, methyl orange, methyl red, resorcin yellow, congo red, chrysoidine, bismark brown etc. Azo dyes are subgrouped into monoazo, diazo, triazo and tetrakisazo dyes according to the number of azo chromophores in the molecule.

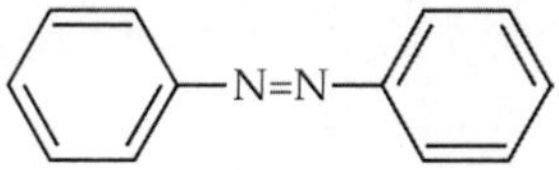

Fig. 5.5: General structure of azo dyes.

Nitro dyes

Nitro dyes contain the nitro chromophore ($–NO_2$). They have the following structure (Fig. 5.6).

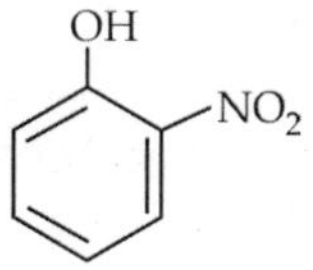

Fig 5.6: General structure of nitro dyes.

The empirical way of designating the nitro group does not indicate that its electrons are delocalised. The bonds to the two oxygen atoms are of equal strength, and are shown as a single and a double bond only for convenience. The electrons from both bonds are shared, and these delocalised electrons are integrated with the delocalised electrons of the aryl rings of dyes, causing absorption in the visible range. The dyes so formed are invariably acid dyes. Picric acid and martius yellow are the usual examples.

PROBLEMS CAUSED BY THE DYES: TOXICITY CONCERN

The toxic effects of synthetic dyes to the environment and human health resulting from the exposure to dyes and dye metabolites are not a new concern. As early as 1895, increased rates in bladder cancer were observed in workers involved in dye manufacturing industries. The effects of occupational exposure to dyestuffs of human workers in dye manufacturing and dye utilizing industries have also received attention. Azo dyes are the largest group of dyes used in the textile industry constituting 60–70% of all dyestuffs produced. They have one or more azo groups (R_1–N= N–R_2) having aromatic rings mostly substituted by sulfonate groups. These complex aromatic substituted structures make conjugated system and are responsible for intense color, high water solubility and resistance to degradation of azo dyes under natural conditions. Color in the effluent is one of the most obvious indicators of water pollution and the discharge of highly colored synthetic dye effluent is aesthetically displeasing and can damage the receiving water body by impeding light. Moreover, azo dyes as well as their breakdown products are cytotoxic or carcinogenic (Khehra *et al.*, 2006). The most acutely toxic dyes for fish are basic dyes, especially those with a triphenylmethane structure. Fish also seem to be relatively sensitive to many acid dyes (Clarke *et al.*, 1980). Over the past decade many microorganisms capable of decolorizing triphenylmethane dyes at lab scale level have been reported (Sharma *et al.*, 2004). The triphenylmethane dye, Basic Violet 3 used as a commercial textile dye, is a recalcitrant molecule indicating that it is poorly metabolized by microbes and consequently is long lived in environment (Chen *et al.*, 2007). Basic violet 3 has been found in water streams as a consequence of improper chemical waste disposal from dyeing industries. Because of its low cost, its effectiveness as an antifungal agent for commercial poultry feed and its ready availability, the general public may be exposed to the dye and its metabolites through the consumption of treated poultry products. Basic Violet 3 is a mutagen, a mitotic poison and clastogens possibly responsible for promoting tumour growth in some species of fish (Cho *et al.*, 2003). The carcinogenic effects of Basic Violet 3 in mice and rodents have also been reported (Decampo and Moreno, 1990). Therefore, there are both environmental and animal health concerns regarding the bioaccumulation of Basic Violet 3.

CONVENTIONAL TECHNIQUES FOR DYE REMOVAL

A wide range of methods have been developed for the removal of synthetic dyes from wastewaters to decrease their impact on the environment (Ramalho, 2004). They are divided in three major categories. The ways of classification of dye removal techniques may include using dye structures, dye classes, or waste stream characteristics but may be of limited use as many methods such as: physical, chemical or biological may be partially effective on a broad range of dyestuffs and dyeing systems. Biological and chemical methods involve the destruction of the dye molecule, whilst physical methods usually transfer the pollutant to another phase.

Physico–Chemical Techniques

Physico–chemical techniques *viz.* membrane filtration, coagulation/ flocculation, precipitation, floatation, adsorption, ion exchange, ultrasonic mineralization, electrolysis, advanced oxidation (chlorination, bleaching, ozonation, Fenton oxidation and photocatalytic oxidation) and chemical reduction are used for the treatment of dye bearing industrial effluents. Oxidative degradation by chlorine and ozone are the most common chemical processes for color removal, but chlorination has the disadvantage of producing organochloride byproducts (Sarasa *et al.*, 1998). Photocatalytic processes are limited to post–treatment units because of the low penetration of UV irradiation in highly colored wastewaters

(Vandevivere *et al.*, 1998). Other physico–chemical technologies, such as membrane filtration or activated carbon adsorption, are expensive and commercially unattractive (Arslan *et al.*, 2002). Table 5.1 summarizes the advantages and disadvantages of physico–chemical techniques due to which they are not seen as permanent solution for the treatment of dye bearing wastewater industries (Robinson *et al.*, 2001).

Table 5.1: Advantages and disadvantages of different physicochemical methods for the treatment of textile dyes (Robinson *et al.*, 2001).

Physical/Chemical methods	*Advantages*	*Disadvantages*
Fenton's reagent	Effective for both soluble and insoluble dyes	Sludge generation
Ozonation	Applied in gaseous state	Short half–life
NaOCl	Initiates and accelerates azo–bond cleavage	Release of aromatic amines
Cucurbituril Electro-	Good sorption capacity	Very expensive
chemical destruction	Breakdown compounds are non–hazardous	High cost of electricity
Activated carbon	Good removal of wide varieties of dyes	Very expensive
Wood chips	Good sorption capacity for dyes	Require long retention time
Silica gel	Effective for basic dye removal	Prevents side reaction
Membrane filtration	Remove all dyes types	Concentrated sludge
Ion exchange	Regeneration: no adsorbent loss	Not effective for all dyes
Irradiation	Effective oxidation at lab scale	Requires lot of dissolved O_2

Biological Techniques

Biological dye removal techniques are based on microbial biotransformation of dyes. Bioremediation is a pollution control technology that uses biological systems to catalyse the degradation or transformation of various toxic chemicals like dyes to less harmful forms.

MECHANISM OF DYE REMOVAL USING YEASTS

The color of the yeast biomass after dye removal is an indicator of the mechanism involved in remediation of dyes. If the yeast biomass appears colorless after dye removal, it suggests that biodegradation of the dye has taken place. If the yeast biomass takes the color of the dye, then the dye removal occurs either through biosorption or bioaccumulation.

Biosorption

Biosorption depends on various parameters *viz.* pH, initial dye concentration and temperature. pH is the most important parameter affecting not only the biosorption capacity, but also the color of the dye solution and the solubility of some dyes. The net charge on biosorbent is also pH dependent because the biosorbent surface has polymers with many different functional groups. Various researchers have investigated the effect of pH on color removal (Aksu and Donmez, 2003; Farah *et al.*, 2007). At lower pH, the biosorbent surface becomes protonated and acquires net positive charge and thus increases the binding of anionic dyes to the sorbent surface. On the other hand, higher pH values increases the net negative charge on the biosorbent surface leading to electrostatic attraction of dye cations. Dye concentration also affects the efficiency of color removal. Initial concentration provides an important driving force to overcome all mass transfer resistances of the dye between the aqueous and solid phases. Hence a higher initial concentration of dye may enhance the adsorption process (Aksu, 2005). The other variable that influences biosorption is temperature. Increase in temperature, decreases the viscosity of the dye solution thereby increasing the diffusion rate of the dye molecules across the external boundary layer and in the internal pores of the adsorbent particles (Farah *et al.*, 2007).

Bustard *et al.* (1998) demonstrated non living biomass derived from *Kluyveromyces marxianus* IMB3 is capable of biosorption of textile dyes *viz.* Remazol Black B, Remazol Turquoise Blue, Remazol Red, Remazol Golden Yellow and Cibacron Orange. It was observed that uptake of Remazol Golden Yellow dye by *K. marxianus* IMB3 was lower at lower concentrations of dye. These results suggested some form of cooperativity with respect to interactions between the dye and the biomass. It was also suggested that the initial interaction between the biomass and the dye resulted in the formation of a layer of dye on the biomass surface. This surface may then became hyper–reactive to other dye molecules, resulting in a stacking effect nucleated by the initial layer of dye on the surface of the biomass. On the other hand, biosorption capacity of same biomass for Cibacron Orange dye increased to a maximum of 8.5 mg/L at a residual dye concentration of 100 mg/L and then decreased rapidly as the equilibrium concentration increased. They suggested one possible reason for this observation that, as the concentration of dye increased above 100 mg/L, dye–dye interactions became prevalent and these interactions might have resulted in decreased affinity of the dye binding sites on the biomass.

Biosorption capacities of nine yeast species (*Saccharomyces cerevisiae, Schizosaccharomyces pombe, Kluyveromyces marxianus, Candida* sp., *C. tropicalis, C. lipolytica, C. utilis, C. quilliermendii* and *C. membranaefaciens*) were studied for removal of Remazol Blue reactive dye from aqueous solutions under laboratory conditions (Aksu and Donmez,2003). The effects of different parameters *viz.* pH, initial dye concentration etc. during the process of biosorption was investigated. Biosorption behavior and biosorption capacities of all the yeasts were different from each other. The yeast *C. lipolytica* showed maximum biosorption capacity at pH 2.0 binding 173.1 mg dye per gram of dry biomass. The Freundlich and Langmuir adsorption models were found to be suitable for describing the biosorption of the Remazol Blue reactive dye by all the *Candida* yeasts (except *C. membranaefaciens*). The biosorption kinetics of nine yeasts for Remazol Blue reactive dye removal at 200 mg/L initial dye concentration for the first 240 min of biosorption was studied. Equilibrium was established in 240 min for these biosorbents for all the initial dye concentrations studied and equilibrium did not change subsequently up to 72 hours. It was suggested that uptake of dye by yeast cells occurred predominantly by surface binding and available sites on the biosorbent are the limiting factors for the biosorption.

Dried biomass of Baker's yeast, *Saccharomyces cerevisiae* was used as biosorbent for Astrazone Blue basic dye from an aqueous solution. The Langmuir model fitted better than the Freundlich and Temkin isotherm model to the experimental data. The calculated heat of adsorption of the dye–yeast system indicated that the biosorption process took place by chemical adsorption and was endothermic in nature (Farah *et al.*, 2007).

Adsorption characteristics of C.I. Reactive Orange 16 (RO 16) from aqueous solution onto residual brewery yeast were investigated under various experimental conditions (Kim *et al.*, 2014). At pH 3.0 and 308 K, the adsorption capacity of RO 16 onto residual brewery yeast was found to be 0.56 mol/kg. Experimental data indicated that the adsorption capacity of residual brewery yeast for the dye was higher in acidic, than in neutral or basic solutions. The adsorption equilibrium data showed good correlation with the Langmuir isotherm models. The estimated values for the free energy of adsorption (ΔG^o) were –6.661, –4.812 and –3.929 kJ/mol at 288, 298 and 308 K, respectively, which indicated that a spontaneous process had occurred. The negative values of enthalpy (ΔH^o) and entropy (ΔS^o) indicated the exothermic nature of the adsorption of RO 16 onto residual brewery yeast. Kinetic studies showed that the biosorption of RO 16 onto residual brewery yeast in the system followed pseudo–second order kinetics.

Mahmoud (2014) reported the potentiality of Baker's yeast strain (*Saccharomyces cerevisiae*) to decolorize a synthetic dye aqueous solution and real industry effluent from Giza spinning and weaving company, Giza, Egypt. The removal of color from one of the azo dyes, Ramazole blue (Vinyl sulfone), had been carried out by Baker's yeast using repeated–batch process. Factors such as solution pH, dye concentration and biomass dosage at different interval times were experimentally tested. The effect of pH on dye bioremoval was investigated at a pH range from 1 ± 0.02 to 6 ± 0.02. The optimum pH values were 2 ± 0.02, 1 ± 0.02, 3 ± 0.02, 4 ± 0.02 and 5 ± 0.02 for direct dye removal, respectively. The effect of dye concentrations was studied using different concentrations of synthetic dye containing 100–600 ppm and the effect of biomass weight was also studied at pH 2 ± 0.02 for different interval times. The equilibrium concentration and the adsorption capacity at equilibrium were determined using two different sorption models namely; Langmuir and Freundlich isotherms. These isothermal models were applied to evaluate differences in the biosorption rates and uptakes of textile dye with a high degree of correlation coefficients in case of Freundlich's isothermal model (R^2 = 0.947). At the end of the experiments, the treatment with Baker's yeast strain could reduce color absorbance and COD value of real textile wastewater by 100% and 61.82%, respectively. Baker's yeast cells were characterized using SEM and FTIR spectroscopy.

Bioaccumulation

Bioaccumulation of reactive textile dyes Remazol Blue, Remazol Black B and Remazol Red RB by *Candida tropicalis* growing in molasses medium was investigated by Donmez (2002). The optimum pH value for bioaccumulation was determined as 3.0 for all three dyes. The maximum specific bioaccumulation capacity of *C. tropicalis* was 111.9 mg/L for Remazol Blue, 101.9 mg/L for Reactive Black and 79.3 mg/L for Reactive Red at approximately 700 mg/L initial dye concentration. The increase in dye concentration inhibited the growth of yeast and caused a long lag phase. Yeast showed highest bioaccumulation percentage of Remazol blue dye among all the dyes tested. The reactive dye bioaccumulation properties of *S. cerevisiae* growing in molasses medium was investigated in a batch system as a function of dye, initial pH and initial dye concentration (Aksu, 2003). The three diazo reactive textile dyes used were Remazol Blue, Remazol Black B and Remazol Red RB. Yeast biomass could provide an effective bioaccumulation for removal of all three dyes at pH 3. The specific dye uptake increased with increasing dye concentration up to 410 mg/L for Remazol Black B, 380 mg/L for Remazol blue and 219 mg/L for Remazol Red RB. Remazol Black B was bioaccumulated much more extensively and faster than Remazol Blue and Remazol Red RB. Remazol Turquoise Blue G (RTBG) Reactive dye bioaccumulation properties of *Candida utilis* has been reported by Gonen and Aksu (2009a). The level of dye accumulation was found to be highly dependent on both initial sucrose and initial dye concentration. *C. utilis* cells showed considerable tolerance to RTBG dye concentrations of < 367.1 mg/L and could accumulate this dye at high quantities.

The bioaccumulation properties of yeast species in case of synthetic dye Basic Violet 3 was reported (Das *et al.*, 2010). Combined effects of sugarcane bagasse extract and synthetic dyes on the growth and bioaccumulation properties of yeasts *P. fermentans* MTCC 189 and *C. tropicalis* showed that the increase in sugar extracted from bagasse played a major role in the growth of yeast and decreased the inhibitory effects of dye on the yeast growth. The growth kinetics of *P. fermentans* MTCC 189 in presence of textiles dyes fitted well with non competitive and uncompetitive inhibition models whereas that of *C. tropicalis* was found to be fitted well with competitive inhibition model. *C. tropicalis* was found to be potential bioaccumulator of dyes than *P. fermentans* MTCC 189. Basic Violet 3 was found to be most toxic for both the yeast species.

Removal of two basic dyes namely Basic Green 4 and Basic Yellow 2 by the yeast *Saccharomyces cerevisiae* was studied in a culture medium prepared from sucrose (Kelewou *et al.*, 2014). The experiments were carried out to determine decolorization. The effects of parameters such as pH, initial dye concentration, the biomass amount and temperature on dye removal were studied. It was found that the decolorization was 96% for the Basic Green 4 (BG4) and 93% for the Basic Yellow 2 (BY2) at pH 5, low initial dye concentration and a temperature of 30°C respectively. Kinetic studies showed that decolorization of Basic Green 4 and Basic Yellow 2 followed a first order kinetic model and bioaccumulation was the process of decolorization.

Biodegradation

Biodegradation is an energy dependent process and involves the breakdown of dye into various by products through the action of various enzymes. When biodegradation is complete, the process is called mineralization (Bennett and Faison, 1997). Biodegradation of crystal violet, a triphenylmethane dye by oxidative red yeasts *viz. Rhodotorula* sp. and *Rhodotorula rubra* was investigated by Kwasiewska (1985). A linear degradation of crystal violet by the yeasts between the second and fourth days of incubation was reported which indicated the presence of an enzyme system for the degradation of crystal violet. It was also observed that fermentative yeast *S. cerevisiae* could not degrade crystal violet.

Biodegradation of an anthraquinone dye, C.I. Disperse Red 15 (DR 15) by yeast strain *Pichia anomala* was reported by Itoh *et al.* (1996). Formation of reaction products *viz.* product 1, PV12.LQ and I–HAQ from DR15 was detected by analytical TLC. The structures of the products are given in Fig. 5.7. Two novel yeast strains *viz. Pseudozyma rugulosa* Y–48 and *Candida krusei* G–1 were reported to be capable of decolorizing multiple types of dyes especially azo types which was largely attributable to biodegradation (Yu and Wen, 2005). Evidence of dye degradation could be judged clearly by inspecting the cell mats. Cell mats became deeply colored because of adsorption of dyes, whereas those retained their original colors did so as a result of biodegradation. Moreover, the disappearance of absorption peaks for all the dyes confirmed that the color removal by *P. rugulosa* Y–48 and *C. krusei* G–1 was largely attributable to biodegradation rather than biosorption onto the yeast cell surfaces. Decolorization of an azo dye, Reactive Black 5 by yeast isolate, *Debaryomyces polymorphus* was reported by Yang *et al.* (2005). It could completely degrade 200 mg/L of non hydrolyzed and hydrolyzed Reactive Black 5 within 24 hours of cultivation. A good correlation was found between dye degradation and the Manganese dependent peroxidase (MnP) enzyme production.

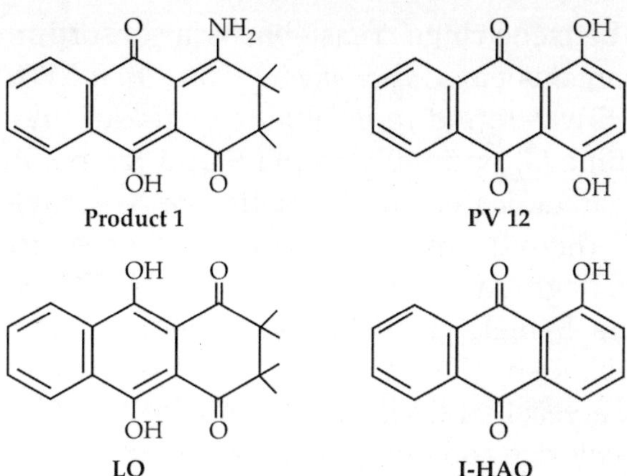

Fig. 5.7: Structure of the reaction products produced after biodegradation of DR 15 by *Pichia anomala* (Itoh *et al.*, 1996).

Candida oleophila, a non conventional ascomycetous wild yeast isolate was identified as an excellent degrader of diazo dye Reactive Black 5 (Lucas *et al.*, 2006). Aerobic batch cultures of *C. oleophila* could completely decolorize up to 200 mg/L in 24 h of incubation at 26°C in presence of glucose (5 g/L). The growth associated decolorization and concomitant absence of extracellular oxidative enzymes in batch cultures indirectly suggested the involvement of an azoreductase–like activity in azo bond cleavage. Reductase enzymes are predominantly involved in biodegradation processes (Olikka *et al.*, 1993; Sani and Banerjee, 1999; Pointing and Vrijmoed, 2000). The effective biodegradation of malachite green, a triphenylmethane dye by *S. cerevisiae* MTCC 463 was reported by Jadhav and Govindwar (2006). A significant increase in the activities of NADH–DCIP reductase and MG reductase was observed in the yeast cells after decolorization which indicated the prominent role of reductases in the degradation processes.

Biodegradation of the Direct Violet 51 azo dye by *Candida albicans* isolated from industrial effluents was reported by Vitor and Corso (2008). Analysis of the metabolites formed after degradation using UV–visible and FT–IR spectroscopy showed breakdown of the dye into various compounds, such as primary and secondary amines and an absence of benzene rings. Yeast–bacteria consortium containing two microorganisms *viz. Galactomyces geotrichum* MTCC 1360 and *Bacillus* sp VUS was found to be able to degrade sulfur containing dye Brilliant Blue G, optimally at pH 9 and temperature at 50°C. GC–MS analysis indicated the formation of 3–{[ethyl–(3–methyl-cyclohexa–2, 5–dienyl)–amino] –methyl}–benzenesulfonic acid as a product of dye degradation by *G. geotrichum* MTCC 1360 alone and (4–benzylidene–3–methyl–cyclohexa 2, 5–dienylidene)–methyl–amine by *Bacillus* sp. (Jadhav *et al.*, 2008b). *Trichosporon beigelii* (NCIM–3326), a yeast strain isolated from the tree *Phoebe porphyria* was able to decolorize and degrade reactive azo dye Navy blue HER effectively into non toxic metabolites (Saratale *et al.*, 2009). Induction in the enzymatic activities of NADH–DCIP reductase and azoreductase presumably indicated the involvement of these enzymes for the biodegradation of Navy blue HER. Biodegradation studies on Basic Violet 3 were conducted using *C. krusei.* Biodegradation was found to be maximum when sucrose was used as cosubstrate (Charumathi and Das, 2011). Enzymes (Lignin peroxidase, Laccase, Tyrosinase, NADH–DCIP reductase, MG reductase, azo reductase) were assayed to find out their involvement in degradation of Basic Violet 3 by *C. krusei* in sugarcane bagasse extract. NADH–DCIP reductase and laccase were found to play dominant role in biodegradation. Based on the results of analysis of intermediate products, the degradation pathway of Basic Violet 3 by *C. krusei* was proposed which occurred *via* stepwise reduction and demethylation process to yield mono–, di–, tri–, tetra–, penta– and hexa–demethylated Basic Violet 3 species and was degraded completely after 24h as shown in Fig. 5.8. Degradation of N, N–dimethyl–N′, N′–dimethyl–N″, N″–dimethyl pararosaniline (Basic Violet 3) occurred *via* stepwise reduction and demethylation processes. The sequence of products formed during reduction was as follows: N, N–dimethyl–N′, N′–dimethyl–N″, N″–methyl pararosaniline, N, N–dimethyl–N′, N′–dimethyl–pararosaniline, N, N–dimethyl–N′, N′–methyl pararosaniline, N, N–dimethyl–pararosaniline, N–methyl pararosaniline and pararosaniline. The compound pararosaniline showed complete degradation.

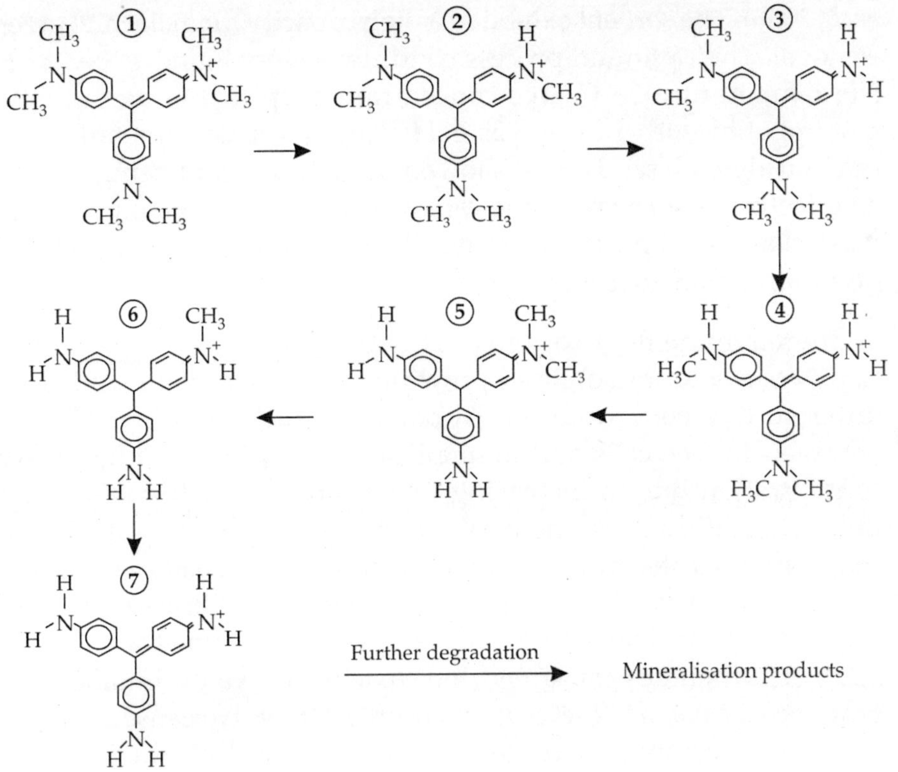

Fig. 5.8: Proposed biodegradation pathway of Basic Violet 3 by *Candida krusei* under shaking condition (Charumathi and Das, 2011).

IMMOBILISATION OF YEAST FOR DYE REMOVAL

It is well known that the free biomass of microorganisms have poor mechanical strength and little rigidity. These factors limit their application in real conditions despite their high dye binding abilities. Immobilization is the possible and practical method for successful reuse of biosorbent in multiple cycles (Vijayaraghavan and Yun, 2007). Compared with freely suspended cells, immobilised microbial system can provide additional advantages, which include efficient and effective regeneration and reuse of the biomass, easier solid–liquid separation and minimal clogging in continuous system. *Saccharomyces cerevisiae* immobilised in calcium alginate–sand beads was tested for green colour removal from aqueous solutions (Godbole *et al.*, 2006). The sorbent exhibited a high capacity for malachite green removal. The optimum process conditions were found to be at pH 5.0, temperature 35°C, dye concentration up to 140 µg/mL and amount of biosorbent was 0.25 g. Hence *S. cervisiae* immobilised on calcium alginate sand beads showed the possibility of being used as an effective adsorbent for the removal of malachite green polluted wastewaters. Reports on immobilised yeast cells used for dye decolourization are seldom known.

To determine the toxic effect of dye Basic Violet 3, free cells of *C. tropicalis* and immobilised growing *C. tropicalis* were studied at different dye concentration. Apparent reduction in growth was observed in free cells and immobilised *C. tropicalis* at higher dye concentration due to increasing toxicity of dye to the yeast cells through inhibition of metabolic activities. However, at all dye concentrations, the growth of yeast in immobilised beads were found to be higher than free cells suggesting immobilisation offered endurance to toxicity of dye on yeast cells (Charumathi and Das, 2010). TEM studies confirmed the toxicity of dye on immobilised growing *C. tropicalis* (Fig. 5.9). Accumulation of dye caused cellular damage and compression of cellular contents which resulted in dome shaped structure of cell (Fig. 5.10).

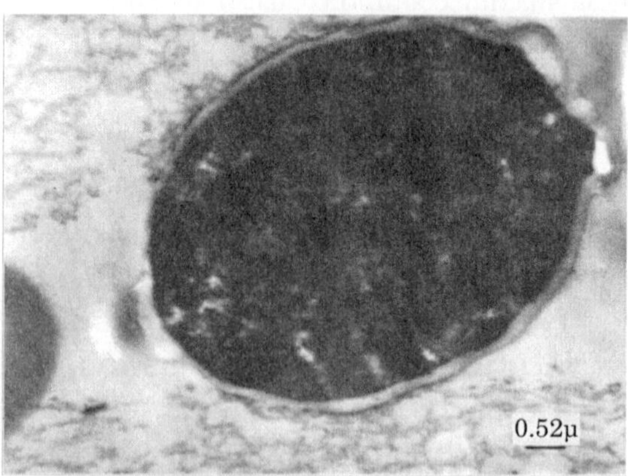

Fig. 5.9: TEM image of yeast cell before bioaccumulation of Basic Violet 3 by immobilised growing *Candida tropicalis*. Initial dye concentration: 50 mg/L; pH: 5.

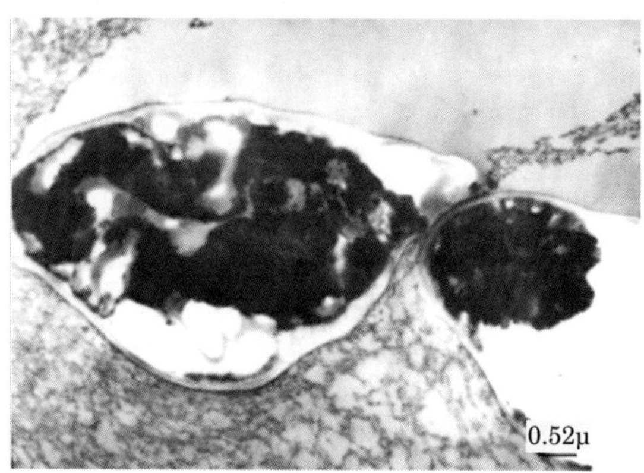

Fig. 5.10: TEM image of yeast cell after bioaccumulation of Basic Violet 3 by immobilised growing *Candida tropicalis*. Initial dye concentration: 50 mg/L; pH: 5.

Charumathi and Das (2012) reported an efficient dye biosorbent which was developed for the treatment of textile wastewater by entrapping dead cells of *C. tropicalis*, within sodium alginate matrix. The biosorbent performance was evaluated in packed bed column with different pH (3 to 6), wastewater strength (25%, 50% 75%), bed height (5 cm–15 cm) and flow rate (0.5 mL/min to 1 mL/min). The conditions *viz.* pH 5, undiluted wastewater, bed height 15 cm and flow rate 0.5 mL min^{-1} were found to be optimum for dye biosorption. The linearized form of the modified Thomas model equation fitted well with the experimental data and described the dynamic adsorption of synthetic dyes from textile wastewater. The Bed depth service time model was used to express the effect of bed height on breakthrough curves. Dye laden immobilised dead *C. tropicalis* was regenerated using 0.01 mol/L NaOH at an elutant flow rate of 1 mL/min. The reusability of the immobilised biomass was tested in consecutive adsorption–desorption cycles.

The FT–IR spectrum of wastewater treated biosorbent showed shifts of several peaks (Fig. 5.12) when compared with the FT–IR spectrum of native biosorbent (Fig. 5.11). However, most significant shifts in the FT–IR spectra were noted from 3433.5 cm^{-1} to 3358.5 cm^{-1} (–NH group superimposed on –OH group), 1661.3 cm^{-1}

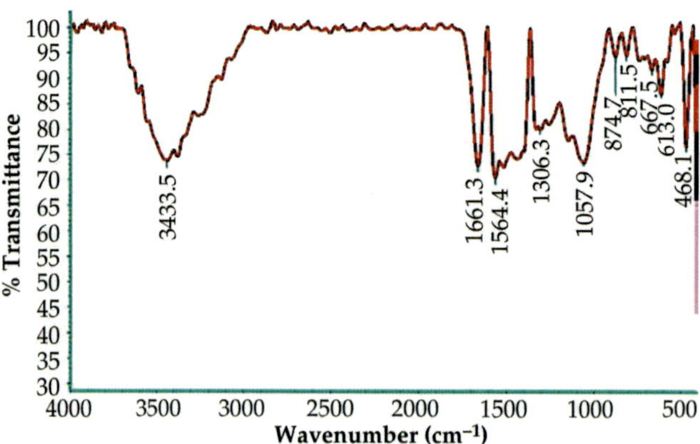

Fig. 5.11: FT–IR spectrum of native immobilised dead *C. tropicalis* before treatment.

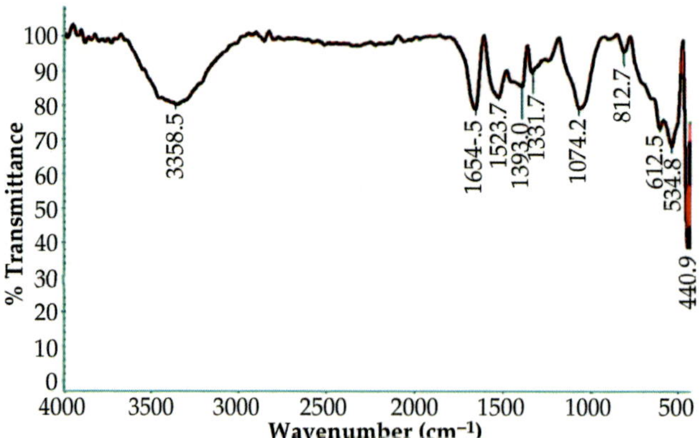

Fig. 5.12: FT–IR spectrum of immobilised dead *C. tropicalis* treated with textile wastewater.

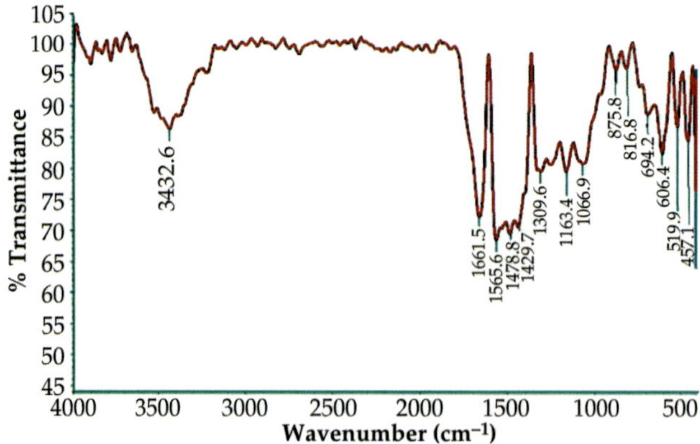

Fig. 5.13: FT–IR spectrum of textile wastewater treated immobilised dead *C. tropicalis* after desorption.

to 1654.5 cm^{-1}, 1306.3 cm^{-1} to 1331.7 cm^{-1}, 1561.4 cm^{-1} to 1523.7 cm^{-1} (–CO conjugated to –NH deformation mode forming the amide group –CO–NH) and 1057.9 cm^{-1} to 1074.2 cm^{-1} (–P–OH group) indicating that amine, hydroxyl, carbonyl, amide and phosphoryl groups were the major sites for sorption of the dyes from wastewater. No major change in the FT–IR spectrum of regenerated biosorbent after desorption (Fig. 5.13) was observed compared to the FT–IR spectrum of native biosorbent which indicated the availability of binding sites for next sorption cycles.

Table 5.2 summarizes the reported information on mechanism of remediation of synthetic dyes using yeast genera.

Table 5.2: Reported information on remediation of synthetic dyes using yeasts.

Yeast	*Dyes*	*Mechanism*	*Reference(s)*
S. cerevisiae	C.I. Reactive Orange 16	Biosorption	Kim *et al.*, 2014
S. cerevisiae	Azo dye, Ramazole blue	Biosorption	Mahmoud, 2014
S. cerevisiae	Basic Green 4 and Basic Yellow 2	Bioaccumulation	Kelewou *et al.*, 2014
Candida tropicalis	Basic Violet 3	Biodegradation	Charumathi and Das, 2012
Candida krusei	Basic Violet 3	Biodegradation	Charumathi and Das, 2011
Candida tropicalis	Basic Violet 3	Bioaccumulation	Das *et al.*, 2011
Candida tropicalis	Basic Violet 3	Biosorption	Charumathi and Das, 2010
Pichia fermentans MTCC 189	Acid Blue 93 Basic Violet 3 Direct Red 28	Bioaccumulation	Das *et al.*, 2010
Trichosporon beigelii NCIM–3326	Navy blue HER Red HE7B Golden yellow 4BD Green HE4BD Orange HE2R Malachite green Crystal violet Methyl violet	Biodegradation	Saratale *et al.*, 2009
Candida albicans	Direct violet 51	Biodegradation	Vitor and Corso, 2008

Table 5.2: (*Contd...*)

Table 5.2: (*Contd...*)

Yeast	*Dyes*	*Mechanism*	*Reference(s)*
Galactomyces geotrichum MTCC 1360	Methyl Red Malachite green	Biodegradation	Jadhav *et al.*, 2008b
Galactomyces geotrichum MTCC 1360 and *Bacillus* sp.	Brilliant Blue G	Biodegradation	Jadhav *et al.*, 2008a
Trichosporon sp.	Vilmafix Blue RR–BB	Biodegradation	Pajot *et al.*, 2007
Candida oleophila	Reactive Black 5	Biodegradation	Lucas *et al.*, 2006
Saccharomyces cerevisiae MTCC 463	Malachite green	Biodegradation	Jadhav *et al.*, 2006
Saccharomyces italicus CICC1201	Reactive Brilliant Red	Biodegradation	Yu and Wen, 2005
Saccharomyces chevalieri CICC1611	Reactive Brilliant Red	Biodegradation	Yu and Wen, 2005
Torulopsis candida CICC1040	Reactive Brilliant Red	Biodegradation	Yu and Wen, 2005
Pseudozyma rugulosa Y–48	Reactive Brilliant Red Weak Acid Brilliant Red B Reactive Black KN–B Acid Mordant Yellow Acid Mordant Light Blue B Acid Mordant Red S–80 Reactive Brilliant Blue X–BR	Biodegradation —	Yu and Wen, 2005
Candida krusei G–1	Reactive Brilliant Red Weak Acid Brilliant Red B Acid Mordant Yellow Acid Mordant Light Blue B	Biodegradation	Yu and Wen, 2005
Issatchenkia occidentalis	Dye I Dye II Dye III Dye IV	Biodegradation	Ramalho, 2004
S.cerevisiae	Remazol Black B Remazol Blue Remazol Red RB	Bioaccumulation	Aksu, 2003

Table 5.2: (*Contd...*)

Table 5.2: (*Contd...*)

Yeast	*Dyes*	*Mechanism*	*Reference(s)*
Debaryomyces polymorphus	Reactive Black 5 Reactive Red M–3BE Procion Scharlach H–E3BE Procion Marine H–EXL Reactive Brilliant Red K–2B Reactive Blue KNR Reactive Yellow M–3R	Biodegradation	Yang *et al.*, 2003
Candida tropicalis	Reactive Black 5 Reactive Red M–3BE Procion Marine H–EXL Reactive Brilliant Red K–2B Reactive Blue KNR Reactive Yellow M–3R	Biodegradation	Yang *et al.*, 2003
Candida zeylanoides	Dye I Dye II Dye III Dye IV	Biodegradation	Ramalho *et al.*, 2002
Candida tropicalis	Remazol Blue Reactive Red Reactive Black	Bioaccumulation	Donmez, 2002
*Kluyveromyces marxianus*IMB3	Remazol Black B Remazol Turquoise Blue Remazol Red Remazol Golden Yellow Cibacron Orange	Biosorption	Bustard *et al.*, 1998
Pichia anomala	Disperse Red 15	Biodegradation	Itoh *et al.*, 1996
Cryptococcus havanensis	Reactive Black 5 Reactive Blue 19	Biosorption	Polman and Breckenridge, 1996
Dekkera bruxellensis	Reactive Black 5 Reactive Blue 19	Biosorption	Polman and Breckenridge, 1996
S.cerevisiae	Reactive Blue 19	Biosorption	Polman and Breckenridge, 1996

Table 5.2: (*Contd...*)

Table 5.2: (*Contd...*)

Yeast	*Dyes*	*Mechanism*	*Reference(s)*
Kluyveromyces waltii	Reactive Black 5 Reactive Blue 19	Biosorption	Polman and Breckenridge, 1996
Pichia casoni	Reactive Black 5	Biosorption	Polman and Breckenridge, 1996
Candida rugosa	Reactive Black 5 Reactive Blue 19	Biosorption	Polman and Breckenridge, 1996
Candida sp.	Procyon Black Procyon Blue Procyon red Procyon Orange	Biosorption	De Angelis and Rodrigues, 1987
Rhodotorula sp.	Crystal violet	Biodegradation	Kwasiewska, 1985
Rhodotorula rubra	Crystal violet	Biodegradation	Kwasiewska, 1985

Dye I=m–[(4–dimethylamino) phenylazo] benzenesulfonic acid
Dye II=p–[(4–dimethylamino) phenylazo] benzenesulfonic acid
Dye III=m–[(2–hydroxy–1–napthyl) azo] benzenesulfonic acid
Dye IV=p–[(2–hydroxy–1–napthyl) azo] benzenesulfonic acid

6

Remediation of Petroleum Hydrocarbon by Yeasts

Petroleum based products are the major sources of energy for industry and daily life. Leaks and accidental spills occur regularly during the exploration, production, refining, transport and storage of petroleum and petroleum products. The amount of natural crude oil seepage was estimated to be 600,000 metric tons per year with a range of uncertainty of 200,000 metric tons per year (Kvenvolden and Cooper, 2003). Release of hydrocarbons into the environment whether accidental or due to human activities is a main cause of water and soil pollution (Holliger *et al.*, 1997).

Petroleum constituents represent: saturates, aromatics, resins and asphaltenes. Structures of some petroleum hydrocarbon constituents are illustrated in the Fig. 6.1. Saturates are defined as hydrocarbons containing no double bonds. They are categorized according to their chemical structures into alkanes (paraffins) and cycloalkanes. Saturates represents the highest percentage of crude oil constituents. Alkanes either a branched or unbranched carbon chains and have the general formula of C_nH_{2n+2} and cyclic alkanes have one or more rings of carbon atoms and general formula of C_nH_{2n}. Aromatic hydrocarbons with one or several aromatic rings are usually substituted with different alkyl groups. In comparison to the saturated and aromatic fractions, the resin and asphaltenes contain non–hydrocarbon polar compounds. Resins and asphaltenes have very complex and mostly unknown carbon structure with addition of many nitrogen, sulfur and oxygen atoms and are more recalcitrant to biodegradation (Harayama, 2004).

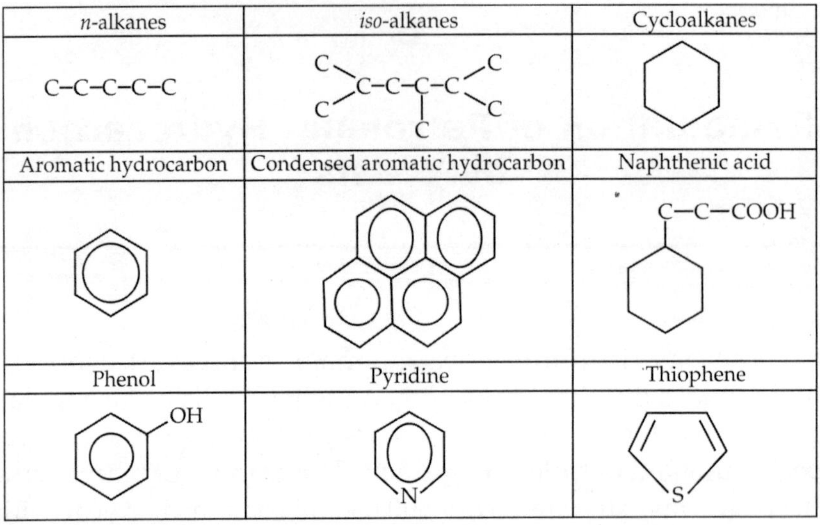

n-alkanes	*iso*-alkanes	Cycloalkanes
C–C–C–C–C	C, C, C–C–C–C, C, C, C	
Aromatic hydrocarbon	Condensed aromatic hydrocarbon	Naphthenic acid
		C—C—COOH
Phenol	Pyridine	Thiophene
OH	N	S

Fig. 6.1: Structural classification of some petroleum hydrocarbon components (Alloway and Ayres, 1993).

Technological development and increasing industrial activities have necessitated a vast increase in the use of diesel oil. It is one of the major products of petroleum industry and has been considered as major pollutant that exerts biohazardous effect on human and other living organisms in the environment (Refaat *et al.*, 2008). Diesel oil is a complex mixture of normal, branched, cyclic alkanes and aromatic compounds with properties of low water solubility, high adsorption coefficient, and high stability of aromatic ring (Van Hamme *et al.*, 2003). There is a considerable risk of environmental hazards during their transport, storage, refining, extraction and disposal. Due to accidental leakage, this product may percolate in the soil and may have an effect on ground water reservoir too. It is frequently reported as oil contaminants leaking from storage tanks and pipelines or released in accidental spills (Gallego *et al.*, 2001). Therefore, pollution caused by diesel oil is a serious environmental concern and it becomes necessary to remove diesel oil from environment. This chapter is specially dealt with the remediation of diesel oil using yeasts as remediation agents.

Existing physical (skimming) and chemical treatments (chemical surfactants or dispersants) for removing environmental contaminants are generally expensive and of limited applicability because sometimes it results in the formation of byproducts that are also harmful than the original. Interests are now being focused on bioremediation technologies for the treatment of the contaminated sites since it is cost–effective and lead to complete mineralization as well as cheaper and effective than the existing conventional methodologies.

Bioremediation functions basically on biodegradation, which may refer to complete mineralization of organic contaminants into carbon dioxide, water, inorganic compounds and cell protein or transformation of complex organic contaminants to other simpler organic compounds by biological agents like microorganisms (Das and Chandran, 2011). One of the important factors that limit biodegradation of oil pollutants in the environment is their limited availability to microorganisms. Petroleum hydrocarbon compounds bind to soil components and they are difficult to be removed or degraded (Barathi and Vasudevan, 2001). Hydrocarbons differ in their susceptibility to microbial attack. The susceptibility of hydrocarbons to microbial degradation can be ranked as follows: linear alkanes > branched alkanes > small aromatics > cyclic alkanes (Ulrici, 2000; Perry, 1984). Some compounds, such as the high molecular weight polycyclic aromatic hydrocarbons (PAHs), may not be degraded at all (Atlas and Bragg, 2009). Microorganisms like bacteria and fungi have been reported to possess the metabolic capacity to use diesel oil as carbon and energy sources for growth and can lead to mineralization. Limited studies have been conducted using yeast as potential degrader of diesel oil.

DIESEL OIL–A MODEL PETROLEUM HYDROCARBON

Diesel oil is a complex mixture of aliphatic and aromatic petroleum hydrocarbons in the range of C_{12}–C_{28}. The boiling point of diesel oil is in the range of 450–670°F. Diesel oil is classified as middle distillate of petrol. The aliphatic alkanes (paraffins) and cycloalkanes (naphthalene's) are hydrogen saturated and contribute 80–90% of diesel. Aromatics (benzene) and olefins (styrene) constitute 10–20% and 1% diesel oils. Petroleum–derived diesel is composed of about 75% saturated hydrocarbons (primarily paraffins including *n, iso,* and cycloparaffins), and 25% aromatic hydrocarbons (including naphthalenes and alkylbenzenes). The average chemical formula for common diesel fuel is $C_{12}H_{23}$, ranging approximately from $C_{10}H_{20}$ to $C_{15}H_{28}$ (ATSDR, 1995). The components of commercial diesel oil are represented in Table 6.1 (Clewell, 1981). The physical and chemical properties of diesel oil are represented in Table 6.2.

Table 6.1: Composition of diesel oil (Clewell, 1981).

Components	*Conc. (% volume)*	*Components*	*Conc. (% volume)*
C10 paraffins	0.9	C14 aromatics	3.8
C10 cycloparaffins	0.6	C15 paraffins	7.4
C10 aromatics	0.4	C15 cycloparaffins	5.5
C11 paraffins	2.3	C15 aromatics	3.2
C11 cycloparaffins	1.7	C16 paraffins	5.8
C11 aromatics	1.0	C16 cycloparaffins	4.4
C12 paraffins	3.8	C16 aromatics	2.5
C12 cycloparaffins	2.8	C17 paraffins	5.5
C12 aromatics	1.6	C17 cycloparaffins	4.1
C13 paraffins	6.4	C17 aromatics	2.4
C13 cycloparaffins	4.8	C18 paraffins	4.3
C13 aromatics	2.8	C18 cycloparaffins	3.2
C14 paraffins	8.8	C18 aromatics	1.8
C14 cyclopraffins	6.6	C19 paraffins	0.7
C19 cycloparaffins	0.6	C19 aromatics	0.3

Table 6.2: Physical and chemical properties of diesel oil.

Characteristics	*Diesel oil*
Molecular weight	Vary
Colour	Colourless to brown
Physical state	Liquid
Melting point	–34°C
Boiling point	193–293°
Odour	Kerosene like
Solubility in water at 20°C	5 mg L^{-1}
Solubility in Organic solvents	Miscible with other petroleum solvents
Vapour pressure at 21°C	2.12–26.4 mm Hg

EFFECTS OF DIESEL OIL ON ENVIRONMENT

Diesel oil contaminates the soil by initiating series of processes affecting both its biotic and abiotic element (Malachowska-Jutsz *et al.*, 1997). This harmful compound changes the fertility of soil. Diesel also has negative effects on biochemical and physiochemical characteristics of soil (Wyszkowska *et al.*, 2002). The soil pollution of diesel leads to the contamination of ground water by percolation through soil and finally reaches on humans. Since diesel is a combination of poisonous toxic chemicals which are carcinogenic, the consumption of polluted water may lead to cancer and toxic chemicals cause skin irritation, indigestion, mental retardation etc. (Marson and William, 2003). Since contamination of soil with refinery products deteriorates its biochemical and physicochemical properties, it also limits the growth and development of plants, whose nutritive and technological value can be low and often questionable. Diesel spills can result in acute toxicity to some forms of aquatic life. Oil coating of birds, sea otters, or other aquatic life which come in direct contact with the spilled oil is another potential hazard.

CONVENTIONAL METHODS FOR REMEDIATION OF PETROLEUM HYDROCARBON CONTAMINATED SOIL

Physical Process

Thermal Treatment

Thermal desorption is an innovative, non incineration technology for treating soil contaminated with organic compound (Fox *et al.*, 1991). It is a proven method in the field of nonhazardous waste treatment and can be used for treating petroleum contaminated soils (Molleron, 1994). Contaminated soil is heated under an inert atmosphere to increase the vapor pressure of the contaminants and for transferring them from the solid to the gaseous phase (Wilbourn *et al.*, 1994). This separates the organics from the soil matrix.

Incineration

For complete destruction of the contaminants, incineration is one of the most effective treatments available. The most common types of incinerators in use are the rotary kiln, multiple hearth, fluidized bed and liquid injection incinerators (Ehrenfeld and Bass, 1984). But incineration may generate incomplete combustion products and a residual ash that may need to be disposed of as a hazardous waste. Moreover, incineration is a relatively expensive process.

Soil Washing

Soil washing is a variation of the soil flushing process, with similar requirements (Lyman *et al.*, 1990). It is performed above ground in a reactor and has been shown to be more effective than the *in situ* flushing system. This approach overcomes some of the problem that may be encountered with *in situ* method *viz.* low hydraulic conductivity, channeling, and contamination of underlying aquifers. There are several state–of–the–art soil–washing systems, including the EPA mobile systems, two water hot systems for removing oil from sandy soils and a floatation process (Assink and Rulkens, 1984). The quantity of residual sludge formed in the extraction process can be a problem and generally requires additional handling as a hazardous waste.

Volatilization

Enhanced volatilization refers to process that removes contaminants from soil by increasing their rate of volatilization (Lyman *et al.*, 1990). This includes the process of mechanical volatilization, enclosed mechanical aeration, pneumatic conveyer systems and low temperature thermal stripping which is considered to be the most effective.

Steam Extraction

Laboratory scale tests and a semi industrial scale plant equipped with vapor condensation and subsequent waste water treatment capability demonstrated that steam extraction can be easily used to remove soil contamination caused by diesel fuel, solvents and polyaromatic hydrocarbons (Hudel *et al.*, 1995). But the technique was found to be costly.

Solidification and Stabilization

This approach incorporates chemical or biological stabilization processes to treat excavated, contaminated soils (Ram *et al.*, 1993). Use of carbon grade fly ash as the only binding agent is a simple, inexpensive method acceptable to the Toxicity Characteristic Leaching Procedure (TCLP) of stabilization and solidification of hazardous wastes like petroleum hydrocarbon.

CHEMICAL PROCESSES

Chemical Treatment

Peroxide spraying can be used to treat excavated, contaminated soil (Ram *et al.*, 1993). Organic substances can be destroyed by indirect electro–oxidation (Leffrang *et al.*, 1995). Commonly used oxidizing agent for the oxidation of petroleum hydrocarbon contaminated soil is cobalt (Co III) because of the high redox potential and the organic carbon in the contaminants are ultimately transformed into carbon dioxide and to small amounts of carbon monoxide which is highly poisonous requiring additional handling as a hazardous waste.

Chemical Extraction

Chemical extraction, such as heap leaching and solid or liquid contactors, can also be used in the treatment of excavated, contaminated soil (Ram *et al.*, 1993). Chemical extraction has been employed on a pilot scale for remediation of soil contaminated with polyaromatic hydrocarbons which are the constituents of diesel oil. But the solvent extraction process is incomplete.

Superficial Fluid Extraction

Oxidation in supercritical water is fast and can lead to total oxidation of the organic compounds in hydrocarbons (Brunner, 1994). Supercritical water is an excellent solvent for extraction of mineral oil fractions from soil, even without oxygen and the effluents are biologically degradable. A supercritical water oxidation system can clean polyaromatic hydrocarbons contaminated soil by extracting hazardous material from the soil and completely destroying it by an oxidation reaction (Lee *et al.*, 1995). Since most organics dissolve readily in supercritical water, the oxidation reaction proceeds very rapidly and produces clean soil. But the process was found to involve high cost and controlled conditions are needed.

BIOLOGICAL PROCESS

Biodegradation

The established physical and chemical remediation methods are inefficient, expensive and can cause recontamination by secondary contaminants. Bioremediation is the microbial degradation of organic pollutants such as petroleum in soil and groundwater. This technique has the benefits of high treatment efficiency, low cost, relatively quick action, *in situ* and *ex situ* application and compatibility with other techniques (Hong *et al.*, 2005). Biodegradation is one of the processes involved in bioremediation which involves natural populations of microorganisms as the primary mechanisms by which petroleum and other hydrocarbon pollutants can be eliminated from the environment (Leahy and Colwell, 1990). Moreover, it is cheaper than other remediation technologies (Ulrici, 2000) and can result in complete mineralization of organic contaminants into carbon dioxide, water, inorganic compounds and cell protein.

BIODEGRADATION OF DIESEL OIL BY YEAST

Compared to bacteria and filamentous fungi, few reports are available on applications of yeasts as potential degrader of diesel oil. Two yeast strains *viz. Saccharomyces cerevisiae* and *Candida albicans* isolated from polluted lagoon water were able to grow effectively on crude oil and diesel as sole sources of carbon and energy (Ilori *et al.*, 2008). They were also found to be the potent producers of biosurfactant. *C. albicans* appeared to be a better diesel–utilizer but the potency of its surfactant production was less than *S. cerevisiae.* Miranda *et al.* (2007) isolated yeast species *Rhodotorula aurantiaca* UFPEDA 845 and *Candida ernobii* UFPEDA 862 from samples polluted by petroleum derivatives collected from the Barra Lagoon near the Suape port terminal, Brazil. In the course of biodegradation assays, *C. ernobii* degraded tetradecane, 5 methyl–octane and octadecane present in diesel oil completely and decane (60.8%) and nonane (21.4%) partially whilst *R. aurantiaca* presented degradation percentages of 93% for decane, 38.4% for nonane and 22.9% for dodecane in diesel over a period of 15 days.

Chandran and Das (2010) isolated five yeast species *viz. Candida tropicalis, Cryptococcus laurentii, Trichosporon asahii, Candida rugosa* and *Rhodotorula mucilaginosa* from petroleum–contaminated soil in India (Fig. 6.2) among which *Trichosporon asahii* was found to be the potent producer of biosurfactant in mineral salt media containing diesel oil as the carbon source and found to be an efficient degrader of diesel oil (95%) over a period of 10 days. In GC–MS analysis, the control system consists of n–alkanes (C9–C26), branched alkanes, naphthalene derivatives, substituted naphthalenes and isoprenoid alkanes (pristine, phytane) (Fig. 6.3), whereas in test system, the yeast species degraded almost all alkanes (Fig. 6.3). Branched chain alkanes and naphthalene derivates were also reduced. The GC–MS analysis suggested that yeast species degraded almost all the hydrocarbon present in diesel (C9%C26), even the isoprenoid compounds like pristine and phytane. Overall percentage degradation of compounds in diesel by *Trichosporon asahii* was found to be 95.01%. Hence this yeast species was considered as suitable bioremediation agent for the remediation of areas contaminated by diesel oil as well as other hydrocarbons. Since diesel oil consists of a variety of molecules such as paraffin, olefins, naphthalene and aromatic compounds, it was considered as a model hydrocarbon to study the hydrocarbon degradation (Ilori *et al.*, 2008).

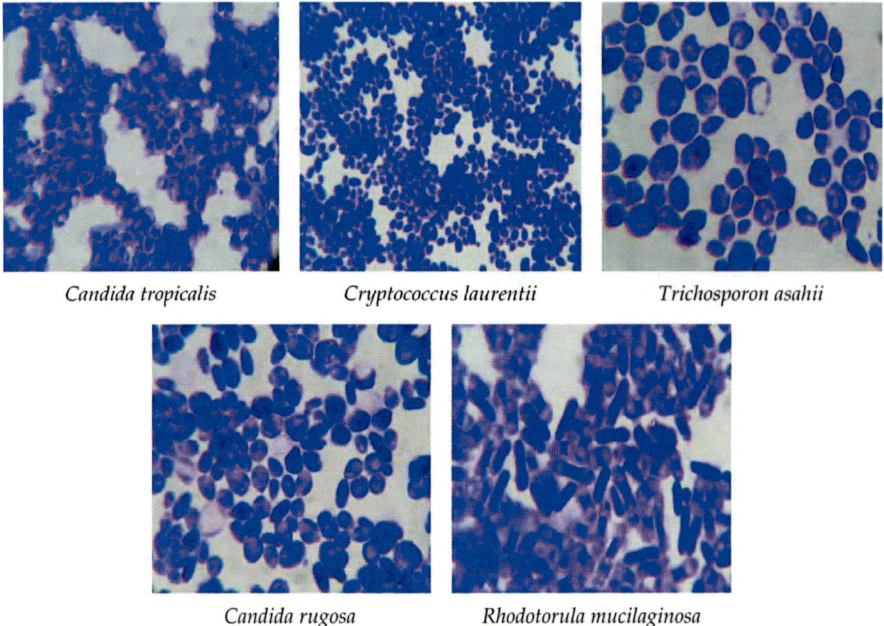

Fig. 6.2: Gram staining images of screened yeast isolates utilizing diesel oil as sole source of carbon and energy.

Production of biosurfactant by *Trichospron asahii* was reported by Chandran and Das (2010). With the use of FTIR spectroscopy, in combination with GC–MS analysis, chemical structures of the purified biosurfactant was identified as sophorolipid species. When compared to synthetic surfactants, including Tween 80 and sodium dodecyl sulfate, the biosurfactant showed high physicochemical

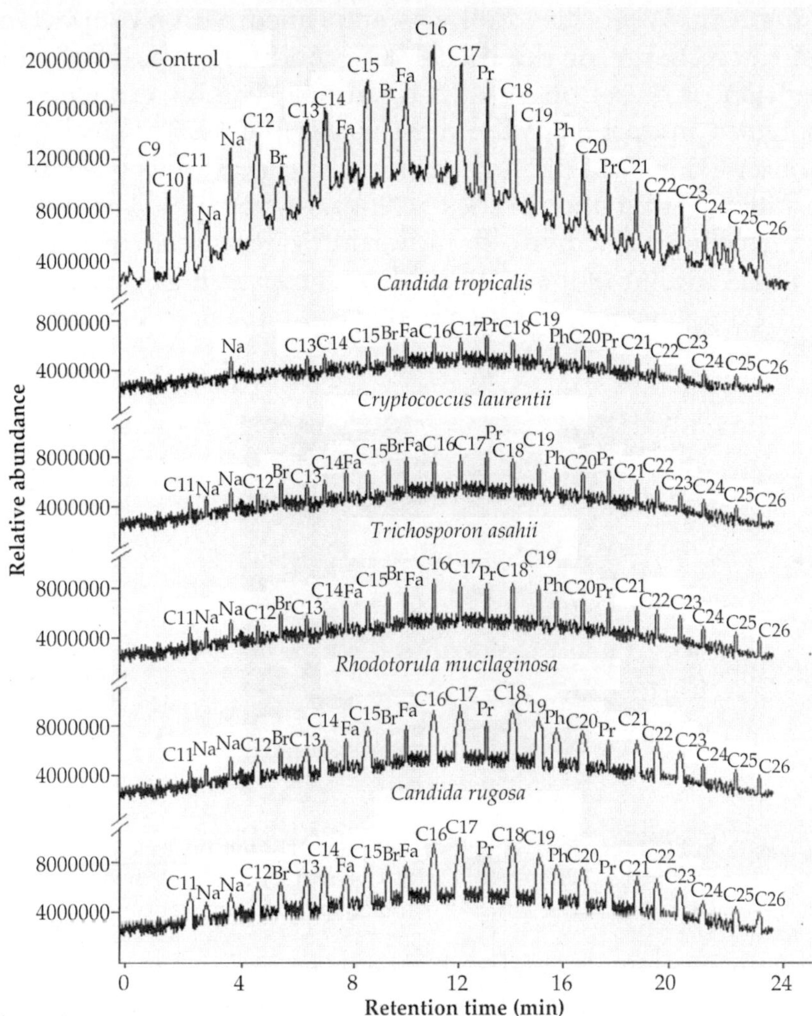

Fig. 6.3: GC peaks of diesel oil before and after degradation by yeast species over a period of 10 days in batch culture period of 14 days. n–alkanes number designates the number of carbon atoms; Br branched alkanes; Na substituted naphthalenes; Fa Farnithane compounds; Pr pristane; Ph phytane (Chandran and Das, 2010).

properties in terms of the surface activities. Involvement of biosurfactant in physiological mechanism of diesel adsorption on yeast cell surface was characterized based on zeta potential measurement. When diesel oil was emulsified with biosurfactant, the surface charge of the diesel was modified resulting more adsorption of diesel on yeast cell surface. The biosurfactant or exopolymer production by the yeast species, *Trichosporon asahii* was well observed in SEM photograph (Fig. 6.4). Yeast cells with diesel oil resulted the formation of sticky colourless matrix of exopolymers interconnecting individual cells into an intricate network of coherent mass after 72 hours of incubation. Microencapsulation of yeast cells in their exopolymers was also noted in SEM image (Chandran and Das, 2010).

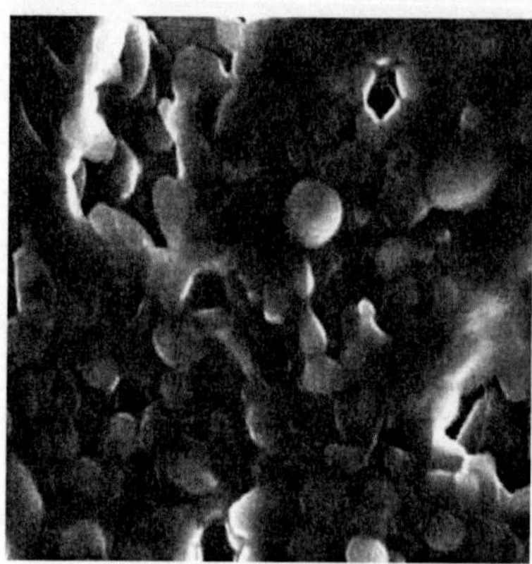

Fig. 6.4: SEM image showing exopolymer production (sticky colorless matrix) by *Trichosporon asahii* grown in Bushnell haas medium containing diesel after 72 h of incubation (Chandran and Das, 2010).

Recently bio–removal of diesel oil has been reported using a mixed microbial consortium consisted of *Rhodotorula aurantiaca* and *Candida ernobii* alongwith other bacterial species isolated from a polluted environment (Silva *et al.*, 2015).

DEGRADATION OF DIESEL OIL BY IMMOBILIZED YEAST AND YEAST BIOFILM

Diesel oil degradation by *Candida tropicalis* immobilized on various conventional matrices (sodium alginate, carboxyl methyl cellulose, chitosan) and biowaste materials (wheat bran, sawdust, peanut hull powder was reported (Fig. 6.5) using the method of entrapment and physical adsorption (Chandran and Das, 2011). The yeast species immobilized in wheat bran showed enhanced efficiency in degrading diesel oil (98%) compared to free cells culture (80%) over a period of 7 days (Fig. 6.6). Copious amount of exopolysaccharides were also produced in the presence of diesel oil. The biofilm forming ability of *C. tropicalis* on PVC strips was evaluated using XTT reduction assay and monitored by SEM and AFM analysis. Yeast biofilm formed on gravels showed 97% degradation of diesel oil over a period of 10 days.

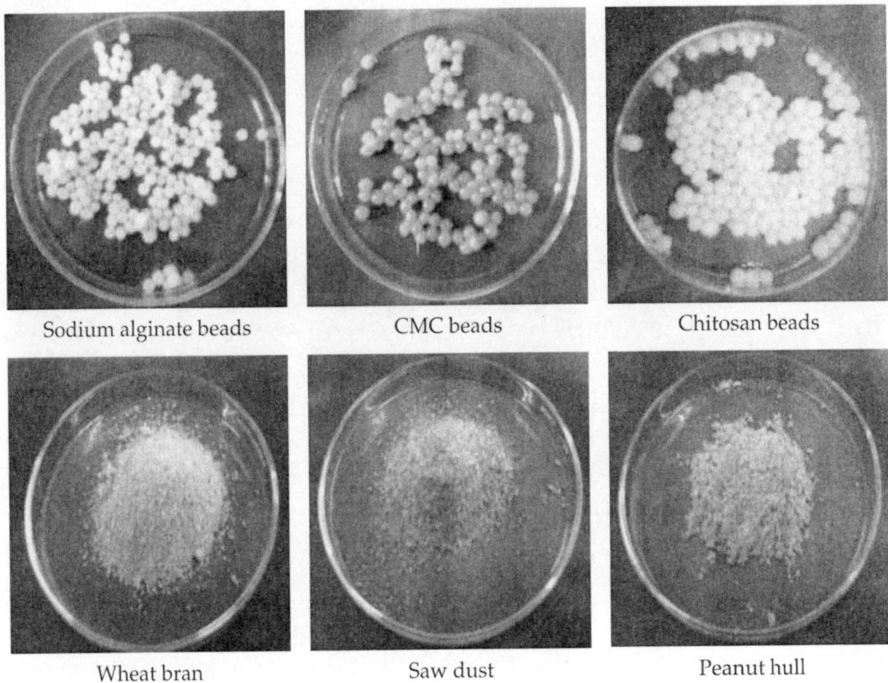

Fig. 6.5: Substrates used for immobilization of yeast species.

ENZYMATIC MECHANISM OF DIESEL DEGRADATION

Five yeast species *viz. Candida tropicalis, Cryptococcus laurentii, Trichosporon asahii, Rhodotorula mucilaginosa* and *Candida rugosa*

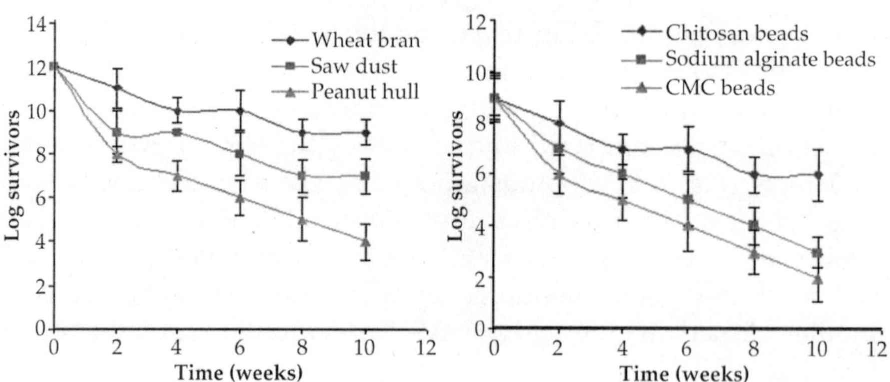

Fig. 6.6: Time–course of the changes in the viable number of *Candida tropicalis* immobilized (a) in biowaste materials (b) in beads during storage at room temperature for a period of 10 weeks (Chandran and Das, 2011).

isolated from hydrocarbon contaminated soil were found to be the potent degraders of diesel oil. These microorganisms showed the presence of enzymes *viz.* Cytochrome P–450, NADPH Cytochrome c reductase, Aminopyrine N demethylase, Alcohol dehydrogenase, Aldehyde dehydrogenase, Naphthalene dioxygenase, Catalase and Glutathione S transferase when the cells were incubated for 48 h in Bushnell haas medium supplemented with 2% diesel oil as sole source of carbon. Cytochrome P–450 monooxygenase enzyme system was found to play an important role in diesel oil degradation. A plasmid of approximately 12 kb size was found to be harboured by all the yeast species (Fig. 6.7). The role of plasmid on diesel oil degradation was assessed by biomass inhibition studies which confirmed that the metabolic machinery of yeast species for diesel oil degradation was plasmid coded. This was the first report establishing the involvement of enzymes and plasmid in diesel oil degradation by yeast species (Chandran and Das, 2012).

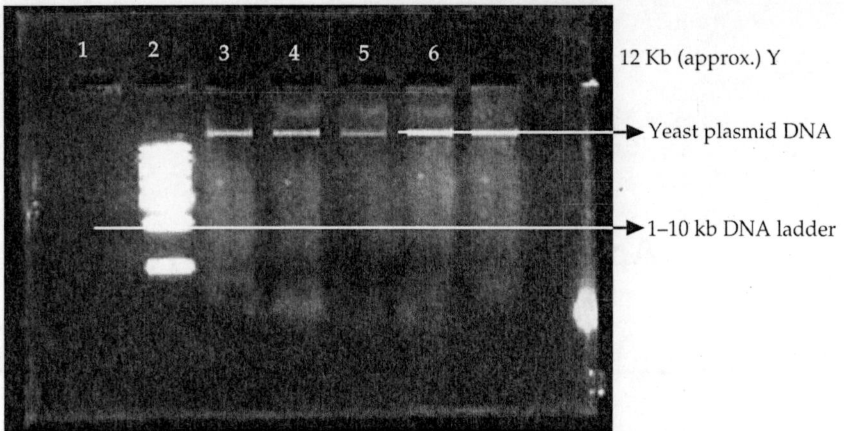

Fig. 6.7: Agar gel electrophoresis of plasmid DNA isolated from diesel degrading yeast species. Lane: 1, 1 Kb DNA ladder (1 Kb to 10 Kb); 2, 3, 4, 5 and 6 shows more than 10 Kb size plasmid DNA from yeasts *Rhodotorula mucilaginosa, Candida rugosa, Trichosporon asahii, Candida tropicalis* and *Cryptococcus laurentii* (Chandran and Das, 2012).

7

Application of Yeasts for Pesticide Removal

Integrated pest management (IPM) is an effective and environmentally sensitive approach to pest management in agricultural settings (USEPA, 2012). In fact, the use of pesticides and bio control agents are two important components of IPM (Talebi *et al.*, 2008). Apart from the beneficial role, the use of pesticides also has some drawbacks, such as potential toxicity to humans and other animals. Organochlorine pesticides are synthetic chemicals used extensively for the control of the diseases of humans and domestic animals that are carried by insects and mites. They are also used against insect pests which cause great damage to agricultural crops (Lal and Saxena, 1982). Stockholm convention on Persistent Organic Pollutants (POPs) is an international environmental treaty, signed in 2001 and effective from May 2004, which aimed to eliminate or restrict the production and use of POPs. In 2004, the Stockholm Convention listed 12 dangerous and persistent organic chemicals as POPs, out of which 9 were pesticides (UNEP, 2005; Gilden *et al.*, 2010). Lindane was added to the list of POPs in the Stockholm Convention conducted in Geneva, along with three other pesticides (MFE, 2011). Production and agricultural use of lindane is banned in 169 countries, but pharmaceutical use is permitted until 2015 (UNEP, 2009). Although most of the developing countries have banned the use of lindane, technical grade HCH is being used extensively in India, even today. Currently, India is the largest consumer and producer of lindane in the world. The continuous use and indiscriminate industrial production, has led to the occurrence of widespread lindane–contaminated soils in the country (Abhilash and Singh, 2010a). Recently lindane concentration in the soil was reported to the range from 28.31 to 31.16 ppm at Sindhuli in Lucknow, India (Vankar and Ramashanker, 2011). Generally lindane residues remaining in the soil enter into the food chain and get concentrated in the fatty tissues of animals causing major health hazards (Benimeli *et al.*, 2007). Therefore, the toxic effect of lindane on the land as well as in the aquatic environment necessitates its removal from the environment. Bioremediation is more efficient and environmentally friendly method for detoxification of lindane residues compared to physical and chemical methods (Zhang *et al.*, 2010a). A number of genera of bacteria, filamentous fungi, actinomycetes and microalgae have been reported with the ability to degrade lindane. This chapter deals with the remediation of lindane, an organochlorine pesticide using yeast species.

LINDANE–AN ORGANOCHLORINE PESTICIDE

Lindane or gamma isomer of hexachlorocyclohexane (γ–HCH) is a highly chlorinated, recalcitrant organochlorine pesticide (OCP), extensively used for the control of agricultural and medical pests (Polti *et al.*, 2014). Lindane exists in eight isomeric forms. It is generally used as an insecticide on fruit, vegetables, forest crops, animals and animal premises. Lindane is highly polar and moderately hydrophobic (Santana–Casiano and Gonzalez–Davila, 1992). It is highly volatile, and when applied to field crops in particular, a high proportion (up to 90%) of the pesticide enters the atmosphere and is later deposited by rain. Half–life for lindane in soil and water is reported as 708 days and 2292 days respectively (Beyer and Matthies, 2001). Substantial exposure to lindane can cause serious health problems like neurological disorders, seizures, convulsions, immunosuppression and even cancer (ATSDR, 2005). Residues of lindane and its isomers have been detected even in the Arctic and Antarctic regions, where they have never been produced or used, due to their ability to migrate *via* various routes (WHO, 2003). Because of its low water solubility, lindane has a strong affinity towards particular matter and consequently reaches to water sediments (Pazou *et al.*, 2006). This compound reaches to aquatic environments through effluent release, atmospheric deposition and water runoff (Yang *et al.*, 2005).

When lindane is applied to a field, several factors control its distribution and fate. Some amount of lindane may be lost to the atmosphere through volatilization, while some portion another may be carried away by surface runoff or photo–degraded by sunlight (Hu *et al.*, 2010). The extent of photodegradation depends on the light intensity and temperature (Malaiyandi *et al.*, 1982) or the presence of photo oxidizers (Malaiyandi *et al.*, 1982; Prakash *et al.*, 1994). There is also evidence that lindane is chemically decomposed under alkaline conditions (El Beit *et al.*, 1981). Early studies with lindane contaminated sites suggested that the degradation was under anoxic conditions and microbial degradation was the primary route of lindane disappearance from soil (MacRae *et al.*, 1967).

The natural degradation of lindane is a very slow process and varies according to the soil type and environmental conditions. In many instances, incomplete breakdown can result in the formation of harmful intermediates (Elcey and Kunhi, 2010). A variety of treatment methods are available for the removal of lindane, which are having some drawbacks (Nagpal and Paknikar, 2006). The physical and chemical methods like removal of lindane by adsorbents, nanoparticles and chemical dechlorination are effective with only lower concentrations of the pollutant. Disposal of the sorbed lindane is another emerging problem (Salam and Das, 2012). Chemical treatment methods which include microwave induced degradation with NaOH modified sepiolite and addition of hydrogen peroxide utilizes corrosive chemicals are not eco–friendly. Physical treatment methods which include thermal desorption and incineration provide efficient degradation, but require huge infrastructure involving high cost. Moreover, they generate highly toxic gases (phosgene). Thus, one way to pacify this problem may be to use bioremediation technology, which is attracting increasing attention as a cost effective and environmentally friendly alternative to the conventional treatment methods (McErlean *et al.*, 2006).

India started the production of technical HCH in the year 1952 (Gupta, 1986) and perhaps was the largest user of technical HCH and DDT in the world. Technical HCH and DDT amounted to 70% of total insecticide production in the 1980's. Around one million tons of technical HCH was used in India during the period 1948 to 1995. Residues of HCH have been reported in soil (Agnihotri *et al.*, 1996; Bhattacharya *et al.*, 2003; Prakash *et al.*, 2004; Singh *et al.*, 2007), drinking water (Mathur *et al.*, 2003; Bakore *et al.*, 2004) food products (Pandit and Sahu, 2002; Kumari *et al.*, 2005; Kumari *et al.*, 2006) and even from soft drinks (Prakash *et al.*, 2004). Realizing the widespread contamination by HCH and toxic nature of HCH isomers, the use of technical HCH was banned in India in 1997 but restricted use of lindane was still permitted (Lal *et al.*, 2008). India has produced 6,353 tons of lindane for export and indigenous use during 1997 to 2006 (Department of Chemicals and Petrochemicals, India) that would mean a production of nearly 60,000 tons of HCH muck consisting of α–, β and δ–HCH. Out of the total, 603.58 tons of lindane has been exported to various countries (Director General of Foreign Trade, New Delhi, India). Nearly 5–8% of lindane was sold to pharmaceutical companies in India, 25–30% was used for formulation and15–20% was exported annually.

PHYSICO–CHEMICAL PROPERTIES OF LINDANE

Lindane is presently used in lotions, creams, and shampoos for the control of lice and mites (scabies) in humans (USFDA, 2009). Trade or other names for lindane include Agrocide, Ambrocide, Aparasin, Aphitiria, Benesan, Benexane, Benhexachlor, Benzene hexachloride, BHC, BoreKil, Borer–Tox, Exagama, Gallogama, Gamaphex, gamma–BHC, Gamma–Col, Gammex, Gammexane, Gamasan, Gexane, Isotox, Jacutin, Kwell, Lindafor, Lindagronox, Lindaterra, Lindatox, Lintox, Lorexane, New Kotol, Noviagam, Quellada, Steward, Streunex, and Tri–6 (USEPA, 1983).

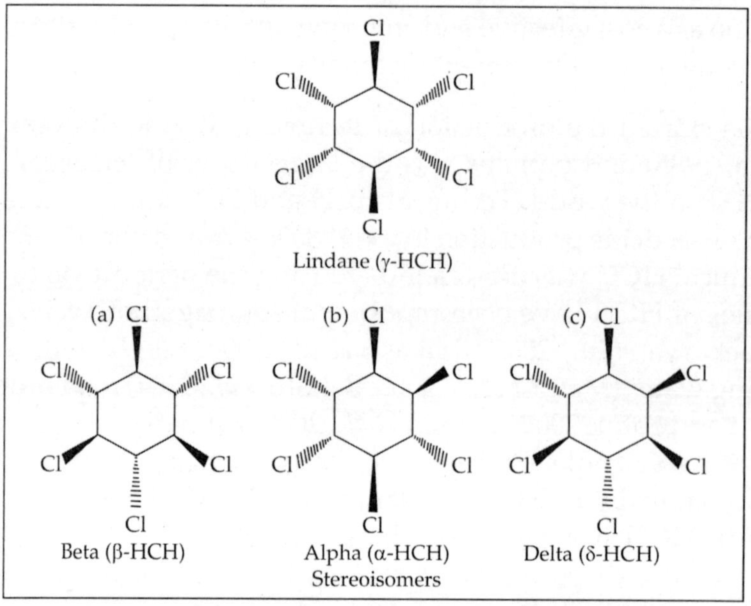

Fig. 7.1: Chemical structure of lindane and major stereo isomers.

Figure 7.1 shows the chemical structure of lindane. It is a white powder, that evaporates in to the air, with a musty (odourless when pure), odour at concentration of 12 ppm and more (NTP, 1998). Lindane is known to be stable in air, light, heat, carbon dioxide, and strong acids. Dehydrochlorination of lindane takes place in the presence of an alkali, forming trichlorobenzene and hydrochloric acid. Dechlorination of lindane occurs on exposure to ultraviolet radiation forming gamma pentachlorocyclohexenes and tetrachlorocyclohexenes (Chalivendra, 2011). The half–life for the environmental degradation of lindane varies depending on the factors such as climate, type of soil, and depth of application. Under humid and field conditions, the half–life of lindane is between a few days to three years (IPCS, 1998). Harner *et al.* (1999) estimated the half–life of lindane in the Arctic oceans as 19 years. Half–life for lindane in soil and water is reported as 708 days and 2292 days respectively (Beyer and Matthies, 2001). Lindane is incompatible with strong bases and powdered metals such as iron, zinc, and aluminium. It is also incompatible with oxidizing agents and can undergo oxidation when it comes in contact with ozone (Chalivendra, 2011). Table 7.1 shows the physic–chemical properties of lindane.

Table 7.1: Physico– chemical properties of lindane.

Common name	*Lindane*
Chemical name	1,2,3,4,5,6–hexachlorocyclohexane, γ–isomer; γ–HCH
CAS Registry number	58–89–9
Chemical formula	$C_6H_6Cl_6$
Molecular weight	290.83
Melting point	112.5°C
Boiling point	323.4°C
Solubility in water at 25°C	7.52 mg/L
Partition coefficients	Log K_{OW}: 3.3, 3.61 Log K_{OC}: 3.0, 3.57
Bioaccumulation factor	in human fat: 19 ± 9 in aquatic animals: 2.5 ± 0.4
Vapor Pressure at 20°C or 25°C	$5.3 \pm 1.4 \times 10^{-3}$ Pa

ENVIRONMENTAL FATE AND CONCERNS

Environmental Protection Agency (EPA) and World Health Organization (WHO) have classified lindane as "moderately hazardous" due to its adverse effects on human health and the environment. The production and agricultural use of lindane are the primary causes of environmental contamination (Abhilash and Singh, 2010). Large amounts of toxic waste in the form of hexachloro-cyclohexane are generated during its production. When lindane is used for agricultural purposes, 12–30% of it volatilizes into the atmosphere which is later re–deposited by rain. Residues of lindane and its isomers have been detected even in the Arctic and Antarctic regions, where they have never been produced or used, due to migration via various routes (WHO, 2003). Because of its low water solubility, lindane has a strong affinity towards particular matter and consequently reaches to water sediments (Pazou *et al.*, 2006).

MODE OF ACTION

Lindane is a contact insecticide which affects the stomach and respiratory system. It acts as a stimulant to the nervous system, causing convulsions and death. In the nervous system, lindane interferes with γ–aminobutyric acid (GABA) neurotransmitter function by interacting with the GABA receptor–chloride channel complex at the picrotoxin binding site (ATSDR, 2005).

TOXICITY OF LINDANE

Lindane is a toxic compound *via* oral exposure. The chief toxic action is on the nervous system where lindane, like other organochlorines, interferes with the flux of cations across nerve cell membranes. Adverse health effects include: apprehension, agitation, mental/motor impairment, excitation vomiting, stomach upset, abdominal pain, central nervous system depression, convulsions, muscle weakness and spasm, loss of balance, grinding of the teeth, hyper–irritability, violent seizures, increased respiratory rate and/or failure, dermatitis, immunotoxicity, and fetotoxicity (IHPA, 2009). The recommended exposure limit (REL) is 0.5 mg/m^3 for lindane. Immediately dangerous to life and health (IDLH) limit is 50 mg/m^3 whereas the maximum contaminant level (MCL) is 0.0002 mg/L for as per the Safe Drinking Water Act.

Toxicity to Humans

Lindane was first introduced as a scabicide for human use in 1950s. It exerts its toxicity mainly by stimulation of the central nervous system (Joy, 1982). Lindane has been associated with numerous severe and fatal adverse reactions. The most serious events and fatalities occurred in pediatric and geriatric populations (Nolan *et al.*, 2012). A direct effect of lindane on striated muscle (rhabdomyolysis) was demonstrated in humans (Munk and Nantel, 1977). At high doses, it may also affect the vascular endothelial lining (disseminated intravascular coagulation) (Sunder *et al.*, 1988). High concentrations of lindane in brain showed neurotoxic effects in humans and several species of animals (ATSDR, 2005). It has been postulated that lindane induces neurotoxicity through several mechanisms, like interfering with γ–aminobutyric acid (GABA), Na^+, K^+–ATPase, Mg^{2+}–ATPase and acetylcholinesterase (AchE) inducing an imbalance of the central monoaminergic systems, alterations in cerebral glucose uptake, induction of brain cytochrome P450s (CYPs) and alteration of neurotransmitter levels (Khan *et al.*, 2013).

Toxicity to Birds

Lindane is toxic to bird species, with a reported LD_{50} of more than 2000 mg/kg in the mallard duck. The 5–day dietary LC_{50} of lindane in Japanese quail is 490 ppm (Hill and Camardese, 1986). The LC_{50} values of lindane in pheasant and bobwhite quail are reported as 561 ppm and 882 ppm, respectively (Ulmann and Blaquiere, 1972). Eggshell thinning and reduced egg production has occurred in birds exposed to lindane (Ulmann and Blaquiere, 1972). Lindane can be stored in the fat of birds. Birds in the Netherlands contained up to 89 ppm in their fat (Ulmann and Blaquiere, 1972). Lindane was reported to be toxic to pigeons causing diarrhea, vomiting, anorexia, depression and sudden death (Blakley, 1982).

Toxicity to Fishes and Aquatic Organisms

Lindane is highly toxic to fish which absorbs it directly from the water or by ingesting contaminated food and bioaccumulate in their fatty acids at ratios of 500:1200 due to its lipophilic nature (Ortiz *et al.*, 2001). Fishes have been used extensively for monitoring purposes because they concentrate pollutants in their tissues, which are directly absorbed from water and also through their diet, reflecting the level of pollution in the aquatic environment (Cazenave *et al.*, 2005). Lindane is highly toxic to fish and aquatic invertebrate species. Reported 96–hour LC_{50} values ranged from 1.7 to 90 µg/L in trout (rainbow, brown, and lake), coho salmon, carp, fathead minnow, bluegill, largemouth bass, and yellow perch (Johnson and Finley, 1980). Water hardness did not seem to alter the toxicity to fish, but increased temperature caused increased toxicity for some species and decreased toxicity for others. The studied LD_{50} in *Etroplus maculatus* was 28 µg/L. Lindane showed profound destructive effects on the gills, liver and kidney of the fish (Nandan and Nimila, 2012).

YEAST AS DEGRADER OF LINDANE

Biodegradation is an efficient and environmentally friendly method for detoxication of lindane residues compared to physical and chemical methods (Zhang *et al.*, 2010a). Salam *et al.* (2013) isolated six yeast strains from three different soil samples collected from the fields of different agricultural plants (Fig. 7.2 and Fig. 7.3). Four yeast strains *viz.*, isolates 1, 2, 3 and 5 (VITJzN01, VITJzN02, VITJzN03 and VITJzN04 respectively) were selected for further studies based on their degradation efficiencies in solid as well as liquid medium and identified at molecular level. The isolate 5, which was identified as *Candida* sp. VITJzN04 showed the best potentiality for lindane degradation in solid as well as liquid media followed by *Rhodotorula* sp. VITJzN03 (Isolate 3), *Pseudozyma* sp. VITJzN01 (Isolate 1) and *Cintractia sorghi* VITJzN02 (Isolate 2). Effect of various parameters *viz.* pH, temperature, shaking speed, inoculum dosage and initial lindane concentration on lindane degradation by *Pseudozyma* VITJzN01, *C. sorghi* VITJzN02, *Rhodotorula* VITJzN03 and *Candida* VITJzN04 were studied in batch experiments. The parameters showed great influence on the growth and utilization of lindane by the yeast strains. At optimum conditions, the yeast strain VITJzN01and VITJzN02 showed 100% degradation of 400 mg/L lindane within 7days, whereas strain VITJzN03 showed 100% degradation of 600 mg/L within 8 days (Salam *et al.*, 2013) and strain VITJzN04 showed 100% degradation of 600 mg/L within 6 days (Salam and Das, 2014). The isolated yeast strains proved to be the most efficient degraders of lindane out of the entire microorganisms reported so far (Salam *et al.*, 2013).

The pH of the medium has a substantial effect on the survival of the microorganisms (Murthy and Manonmani, 2007). The growth of the yeast strains were monitored at a wide range of pH, 3.0–8.0 in MM supplemented with 100 mg/L lindane and lindane utilization

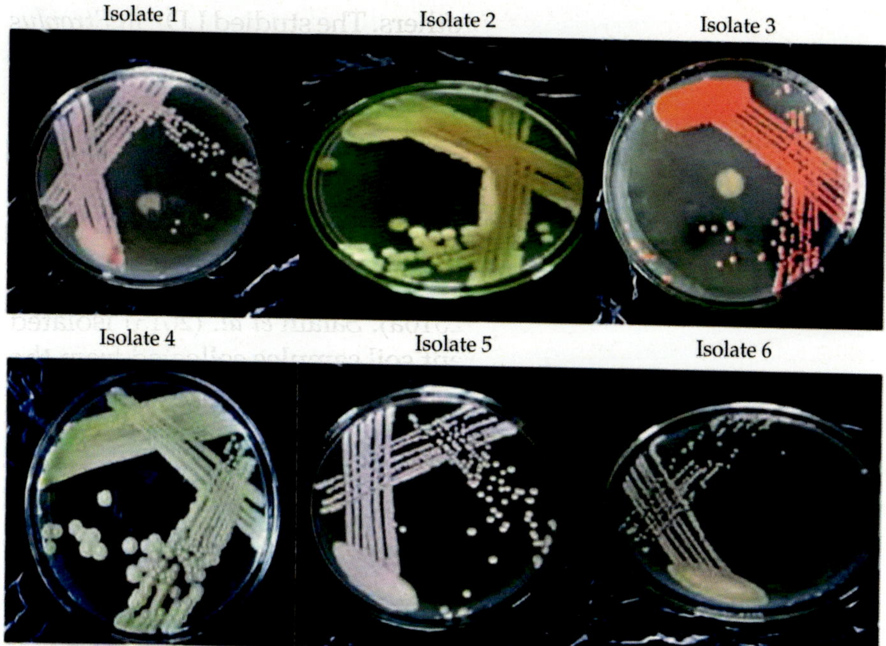

Fig. 7.2: Colonies of the isolated yeasts on YEPD agar plates after 48 h.

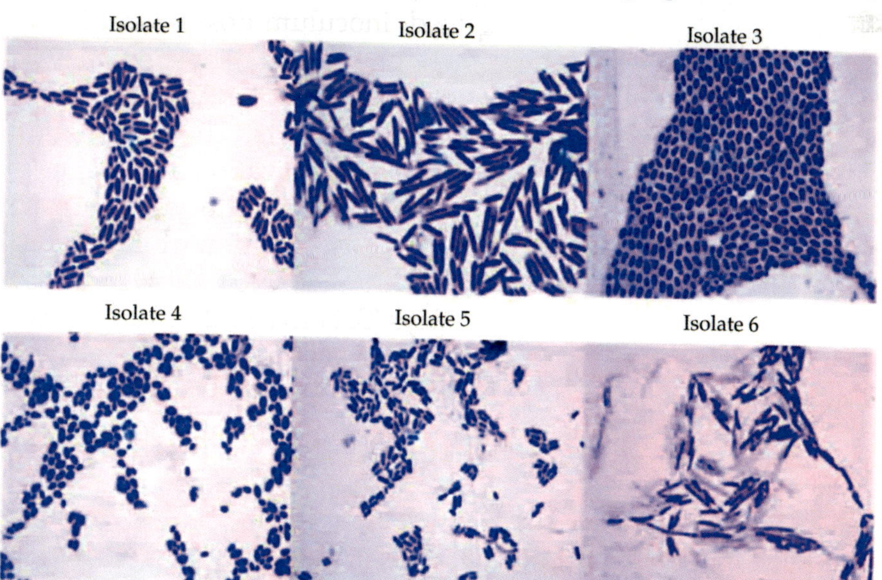

Fig. 7.3: Microscopic images of the yeast isolates at 100X objective of the Bright field microscope.

by the yeast isolates was best seen under pH ranging from 6.0–8.0. (Salam *et al.*, 2013; Salam and Das, 2014). Previous reports suggested that the degradation of lindane was most favourable under a neutral range of pH 6.0–8.0 (Manonmani *et al.*, 2000; Siddique *et al.*, 2002; Murthy and Manonmani 2007; Elcey and Kunhi, 2010). To study the effect of agitation on lindane degradation, the yeast cultures were incubated at different shaking speeds (60–140 rpm). Cultures incubated at a shaking speed of 120 rpm showed the best removal of lindane after 10 days (Salam *et al.*, 2013; Salam and Das, 2014) Growing cultures in liquid medium with proper agitation provides uniform distribution of nutrients, oxygen and microorganisms throughout the process (Elcey and Kunhi, 2010). Lindane utilization was increased with the increase in inoculum load to certain limit after which further increasing the number initial inoculum did not show any effect in the cell dry weight yield as well as degradation of lindane (Salam *et al.*, 2013; Salam and Das, 2014). This may be due to the attainment of an equilibrated level of inoculum over the incubation period.

The concentration of the target pollutant is an important factor which affects the biomass production as well as the degradation rates. Low concentration of pollutant might not be able to induce the enzymes of degradative activity while high levels may be toxic to the microorganisms (Awasthi *et al.*, 2000). Initial concentration of 400 mg/L of lindane was found to be optimum for VITJzN01 and VITJzN02 whereas a higher concentration of 600 mg/L was optimum for the growth of VITJzN03 and VITJzN04 and complete removal of lindane. Concentrations above 600 mg/L were found to show inhibitory effects on yeast growth. Several microorganisms capable of degrading lindane were reported earlier. But this was the first report on complete degradation of lindane at a high concentration such as 600 mg/L by yeasts *Candida* VITJzN04 within shortest period of 6 days (Salam and Das, 2014).

ENZYMATIC MECHANISM OF LINDANE DEGRADATION

Oxidative enzymes and cytochrome p450 enzymes have been reported to be responsible for pesticide degradation in eukaryotes

(Van Eerd *et al.*, 2003). Studies have focused mostly on the activities of the enzymes *viz.* laccase, Mn peroxidase and lignin peroxidase to understand the role of fungi in lindane degradation (Nagpal *et al.*, 2008; Guillén–Jiméneza *et al.*, 2012). Significant enzyme activities *viz.* Mn–peroxidase and lignin peroxidase along with lindane dechlorinase, lindane dehalogenase, catechol–1,2–dioxygenase and 2,5–dihydroquinone reductive dechlorinase were reported during lindane degradation over the incubation period. In contrast, mild laccase activity was seen in the cultures of *Pseudozyma* VITJzN01 but absent in all other yeasts studied. This was the first report on the involvement of the enzymes *viz.* lindane dechlorinase and lindane dehalogenase during lindane degradation in yeast species (Salam and Das, 2014). GC–MS–based metabolite analysis has profound implications for discovering the mode of action of pesticides. It helps to unravel the effect of altered gene expression on metabolism and organism performance in biotechnological applications (Du *et al.*, 2013) Most common metabolites reported for aerobic degradation of lindane are γ–PCCH, 2, 5–dichlorobenzoquinone, chlorohydroquinone, chlorophenol, and phenol. Whereas PCCH, isomers of trichlorobenzene (TCB), chlorobenzene and benzene are the most typical metabolites found in anaerobic pathways (Camacho–Perez *et al.*, 2012). Salam *et al.*(2013) reported seven intermediates including c-pentachlorocyclohexane (c–PCCH), 1,3,4,6–tetrachloro–1,4–cyclohexadiene(1,4–TCCHdiene), 1,2,4–trichlorobenzene (1,2,4 TCB), 1,4–dichlorobenzene (1,4 DCB), chloro–cis– 1,2–dihydroxycyclohexadiene (CDCHdiene), 3–chlorocatechol (3–CC) and maleylacetate (MA) derivatives indicating that lindane degradation follows successive dechlorination and oxido–reduction which were formed at different periods of growth during the process of lindane degradation by *Rhodotorula* VITJzN03. In case of the strain *Candida* VITJzN04, a reductive dechlorination pathway was observed after the initial dechlorination of lindane to form γ–PCCH. The parent lindane was first dechlorinated to form penta–chloro and subsequently the tetra–chloro derivative. Further by multi step reduction and dechlorination, 2,5–DCHQ was formed and reduced to 2–CHQ. The closed ring was then opened and hydrolysed to produce HMSA.

HMSA was oxidized to form maleyl acetate which was the last metabolite detected in GC and was incorporated directly into cellular tricarboxylic acid cycle for generation of energy. Some intermediates such as dichlorocyclohexa–2,5–diene–1,4–diol (DCCH–2,5–diene–1,4–diol) and hydroquinone (HQ) are shown with doted–lines. These compounds might have formed during lindane degradation in minute quantities and was spontaneously transformed, which were not detected in gas chromatogram (Salam and Das, 2014) as shown in Fig. 7.4.

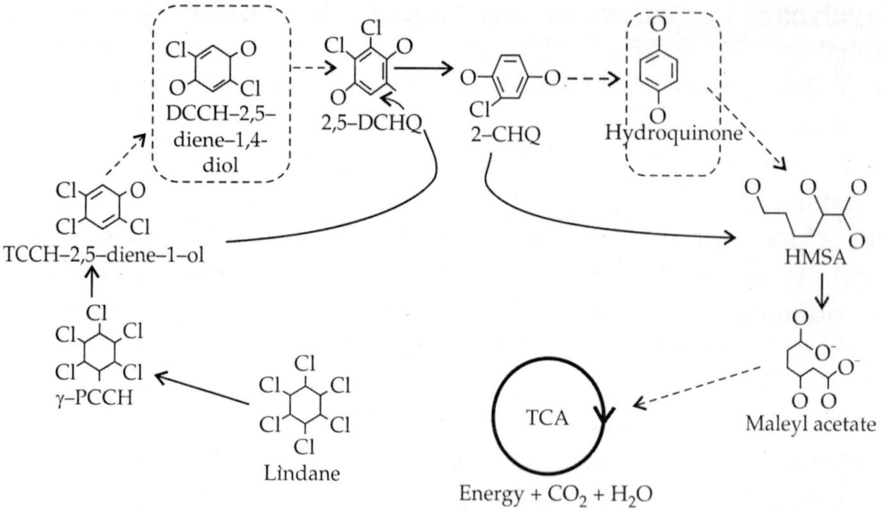

Fig. 7.4: Proposed degradation pathway of lindane degradation by the isolated yeast strains *Candida* VITJzN04 (Salam and Das, 2014).

Enzyme analysis and degradation product characterization in the yeast isolates provided a strong foundation for further research in the field of lindane remediation and also for the implementation of a powerful bioremediation strategy for the decontamination of lindane contaminated environments.

BIOSURFACTANT PRODUCTION BY YEAST STRAINS

Production of biosurfactants such as BS–I, II, III and IV by the four isolated yeast strains *viz. Pseudozyma* VITJzN01, *C. sorghi* VITJzN02,

Rhodotorula VITJzN03 and *Candida* VITJzN04 was reported and structural characterization was done following FTIR, NMR and GC–MS analysis. According to the structural determination, the BS–I, II, III and IV were identified as mannosylerythritol lipid–A (Salam and Das, 2013a), mannosylerythritol lipid–B, lactonic di–acetyl sophorolipid (Salam and Das, 2013b) and acidic di–acetyl sophorolipid respectively. Figure 7.5 shows the structures of the biosurfactants produced by yeast strains.

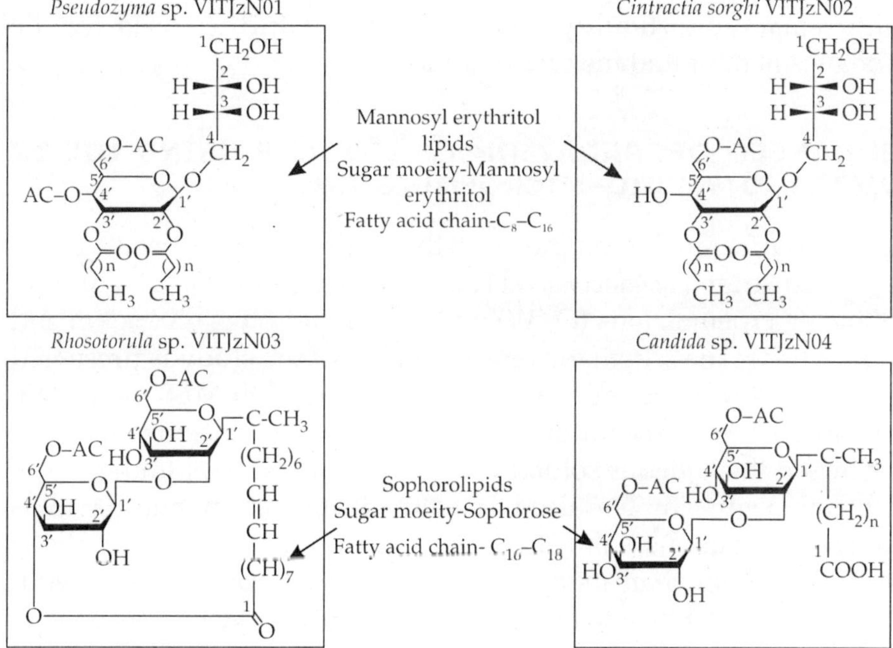

Fig. 7.5: The structures of the biosurfactants produced by the lindane degrading yeast strains (Salam and Das, 2013a).

Biosurfactant production by yeast strains were also reported by other workers. NMR analysis of glycolipid produced by yeast strains belonging to the genera *Pseudozyma* and *Ustilago* isolated from sugarcane were reported to produce mannosylerythritol lipids from sugarcane juice (Morita *et al.*, 2012). *Pseudozyma rugulosa* NBRC 10877 also produced similar MEL glycolipids (Morita *et al.*, 2006). Yeasts belonging to *Pseudozyma* sp. are well known producers of glycolipids from various substrates like soyabean oil (Morita *et al.*, 2006a,b; Konishi *et al.*, 2007) and n–alkanes (Kitamoto *et al.*, 2001). A strain of *Candida* sp SY16 was also reported to produce glycolipid homologous to MELb (Kim *et al.*, 1999). Another yeast *Candida batistae* was found to produce a novel acid sophorolipid from olive oil (Konishi *et al.*, 2008). In a recent study, a strain of *Rhodotorula mucilaginosa* KUGPP–1 was found to produce a novel BS composed of glycoproteins (Kawahara *et al.*, 2013). The biosurfactant MELs produced by the yeast strain VITJzN01 showed the lowest CMC value and remarkable stability. Therefore, it was further used for the enhancement of lindane degradation.

ENHANCED DEGRADATION OF LINDANE USING OIL IN WATER (O/W) BIO–MICROEMULSION

The biosurfactant BS mannosylerythritol lipids (MEL) produced by the yeast strain *Pseudozyma* VITJzN01 was used for the preparation of bio–microemulsions (O/W) with water and olive oil (Salam and Das, 2013a). The schematic representation of the study is presented in Fig. 7.6 O/W bio–microemulsions formed with MEL effectively enhanced the degradation of lindane at high concentration and increased the aqueous solubility and bioavailability of lindane. The bio–microemulsion was used for solubilization of lindane 100 folds above its aqueous solubility limit and lindane was solubilised 40 folds higher than its aqueous solubility. Applications of MEL stabilized O/W bio–microemulsions without cosurfactant for lindane degradation was the first report of eco–friendly approach over the use of conventional surfactant micelles to enhance the bioremediation of lindane in soil and aqueous environment (Salam and Das, 2013a).

Manickam *et al.* (2012) reported that the aqueous solubility of lindane was enhanced 11 fold by the addition of 40–60 μg/mL concentration of biosurfactant sophorolipid in a culture of *Sphingomonas* sp. NM05. Sharma *et al.* (2009) reported that BS produced by *Pseudomonas aeruginosa* WH–2 improved the aqueous phase partitioning of lindane by 70%. Zheng *et al.* (2012a) reported the increase of degradation of DDT by fungi *Phanerochaete chrysosporium* at high concentration with the use of microemulsions formed with non–ionic surfactants, linseed oil and co–surfactant.

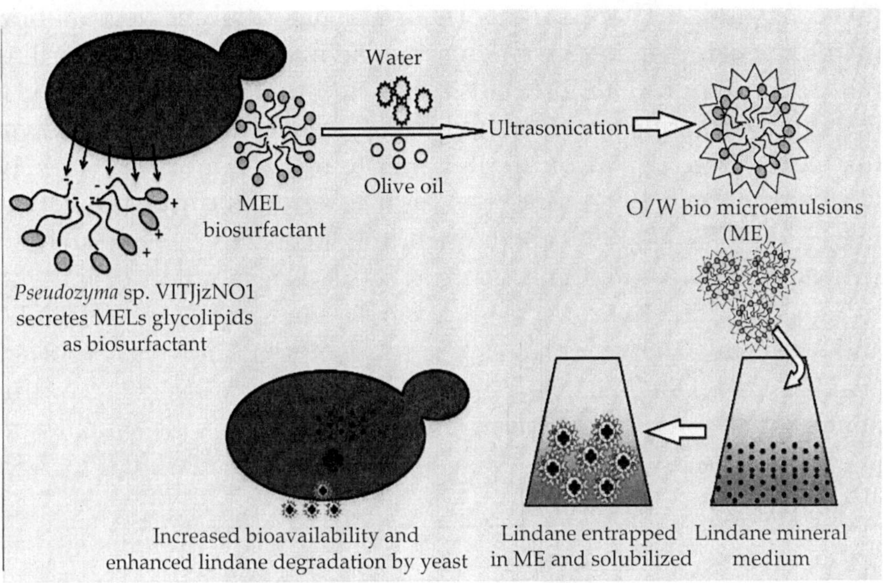

Fig. 7.6: Schematic representation of application of biomicroemulsion for enhanced degradation of lindane. The yeast *Pseudozyma* sp. VITJzN01 produced biosurfactant MEL with hydrophilic (+) and hydrophobic (–) moieties. O/W bio–microemulsions Bio–(ME) were formed with MEL, olive oil and water by ultra sonication. The O/W Bio–ME was used to solubilize lindane in mineral medium for increased bioavailability and degradation by the yeast in the liquid medium (Salam and Das, 2013a).

LINDANE DEGRADATION USING YEAST MEDIATED BIO–NANO HYBRID SYSTEM

In recent years, a new approach toward degradation of environmental contaminants was proposed by few researchers which involve sequential or simultaneous treatment of target contaminants. A concomitant/sequential technique involves the utilization of biotic (microorganism) and abiotic (metal catalysts) process mutually in an integrated method to degrade the target contaminant in a better way (Singh *et al.*, 2013). These hybrid processes possess several advantages both environmentally and commercially such as: (1) milder conditions, (2) elimination of toxic and expensive reagents, (3) ease of catalyst/product separation, (4) good selectivity and yields, (5) eco–friendly. There are reports on the bio–catalytic degradation of lindane using nano zero valent iron and nano palladium (Mertens *et al.*, 2007; Singh *et al.*, 2013). It is reported that zinc oxide can be a suitable alternative to zero valent iron since it is an efficient catalyst for water detoxification as it generates H_2O_2 more efficiently (Carraway *et al.*, 1994). ZnO has high reaction and mineralization rates (Poulios *et al.*, 1999) and more numbers of active sites with high surface reactivity (Pall and Sharon, 2002). Degradation of lindane using a bio–nano hybrid system composed of *Candida* sp. VITJzN04 and nano scale zinc oxide (n–ZnO) in aqueous solution was reported (Salam and Das, 2015). Yeast cells coated with zinc oxide nanoparticles were termed as nano–decorated yeast cells which were used as a bio–nano hybrid system for lindane degradation (Fig. 7.7).

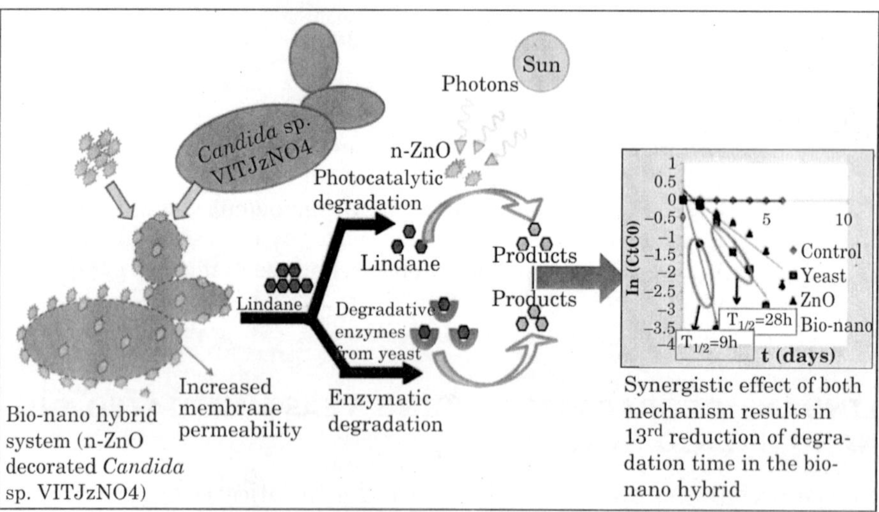

Fig. 7.7: Schematic representation of the application of bio–nano hybrid system for lindane degradation.

n–ZnO particles were used for coating the yeast cells to form a bio–nano hybrid system for lindane dechlorination and degradation. This was the first report on the application of n–ZnO for photocatalytic removal of lindane. Concentration of ZnO nanoparticles for decoration of the yeast was optimized. The yeast *Candida* sp. VITJzN04 showed greater resistance even to a concentration of 50 mg/mL of the ZnO nanoparticles. The decorated yeast cells showed complete degradation of lindane in 9 h which is a remarkable decrease in time compared to the results reported so far. Further, major enzymes responsible for lindane degradation in *Candida* sp. VITJzN04 were analyzed. The experiments revealed that the activity of enzymes were increased almost 5 folds when n–ZnO was coated on yeast cells. The enhanced activity is possibly due to immobilization of the degradative enzymes onto n–ZnO both in the extracellular and intracellular matrix. The enhanced activity of enzymes immobilized on nanoparticles was also reported by Dyal *et al.* (2003) and Song *et al.* (2012). Such immobilized enzymes have better stability than naked enzymes which can in turn results in the rapid and effective degradation. Lindane degradation might have occurred through an integrated approach of bio–nano hybrid system involving photocatalytic degradation of lindane using n–ZnO and biodegradation of lindane by *Candida* VITJzN04 in aqueous medium which may serve as an effective remediation tool for the treatment of wastewater and ground waters containing lindane.

Scanning electron micrographs of the surfaces of the native cells and yeast cells decorated with n–ZnO are shown in Fig. 7.8. The native cells (A) had normal ovoid cell morphology, whereas, the decorated cells (B and C) clearly showed that the n–ZnO particles were adsorbed on the surface of the yeast. In Fig. 7.8B, the magnified image of yeast cell decorated with n–ZnO showed alteration in the cell morphology due to intracellular accumulation of the nanoparticles. In this condition also the yeast cells were live and bud formation was noted in the image. In Fig. 7.8C, along with budding, the surface pore and cracks could be visualized. Fig. 7.8D shows the yeast cells were recovered from the stationary phase of lindane degradation. Compared to the native yeast cells, clear shrinkage and cracks could

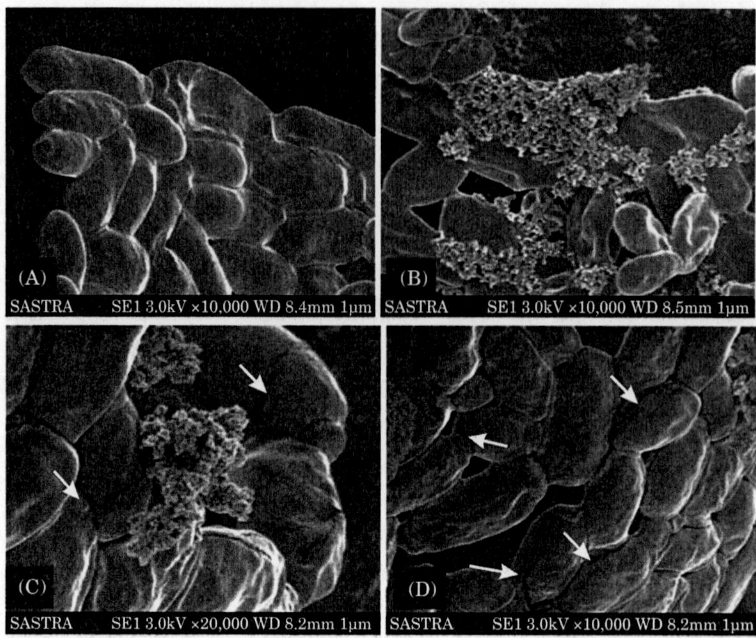

Fig. 7.8: SEM images of *Candida* sp. VITJzN04. (A) native yeast cells (B) Bio–nano hybrid–n–ZnO decorated yeast cells (C) decorated yeast cells showing cracks on the cell surface due to n–ZnO interaction (D) Lindane interacted bio–nano hybrid cells recovered from stationary phase (Salam and Das, 2015).

be visualized in these images. These cracks formed due to the accumulation of nanoparticles in the yeast cells which might be responsible for easy transport of lindane into the intracellular matrix for enhanced degradation of the compound.

To understand the reaction kinetics of lindane removal, the degradation data were fitted to the first order model (Fig. 7.9B). The degradation rate constant and half–life was calculated from the first order kinetic plot and presented in Table 7.2. The half–life of lindane degradation was 1.17 days (almost equal to 28 h) for the yeast *Candida* sp. VITJzN04 which was further reduced in the case of the bio–nano hybrid system being 0.39 days (almost equals to 9 h). The results suggested that the degradation of lindane by the bio–nano hybrid system proved to be more efficient than the individual systems which could bring down the half–life to 1/3rd of the value as shown by the individual yeast (Salam and Das, 2015).

Table 7.2: Comparison of the kinetic parameters during lindane degradation by various treatments.

Treatments	*Rate Constant; K (per day)*	*Half–life; $T_{1/2}$ (day)*
Control	0.0009	770
n–ZnO	0.354	1.95 (46 h)
Candida VITJzN04	0.589	1.17 (28 h)
Bio–nano hybrid	1.75	0.39 (9 h)

BIOSORPTIVE REMOVAL OF LINDANE

Biosorption as a means of remediation has emerged as an efficient, cost–effective and environmental friendly technique alternative to the conventional techniques for remediation of various organic pollutants (Wu and Yu, 2007). The use of non–living biomaterials as sorbent have the advantage of not requiring utmost care and maintenance as well as being useful in remediating areas with high level contamination (Manzoor *et al.*, 2013). There are few reports on microbial species including bacteria and fungi which are capable of removing lindane by biosorption showing low uptake capacity (Tsezos and Bell, 1989; Ju *et al.*, 1997).

Lindane biosorption capacity of native yeast biomass of *C. sorghi* VITJzN02 and *Rhodotorula* VITJzN03 was reported which was found to be maximum at pH 3.0 (Salam and Das, 2013c). Lindane has a weak negative charge with spacial arrangement of Cl^- atom around it. As a charged species, the degree of its sorption onto the biosorbent surface was primarily influenced by the surface charge on the biosorbent, which in turn was influenced by the solution pH. At acidic pH, the protonation of surface functional groups on the biosorbent took place, which enhanced the approach of negatively charged lindane molecule to the surface of the biosorbent resulting increased biosorption. Similar observations were earlier reported by Ju *et al.* (1997), Gupta *et al.* (2002) and Hassan *et al.* (2009). The sorption data

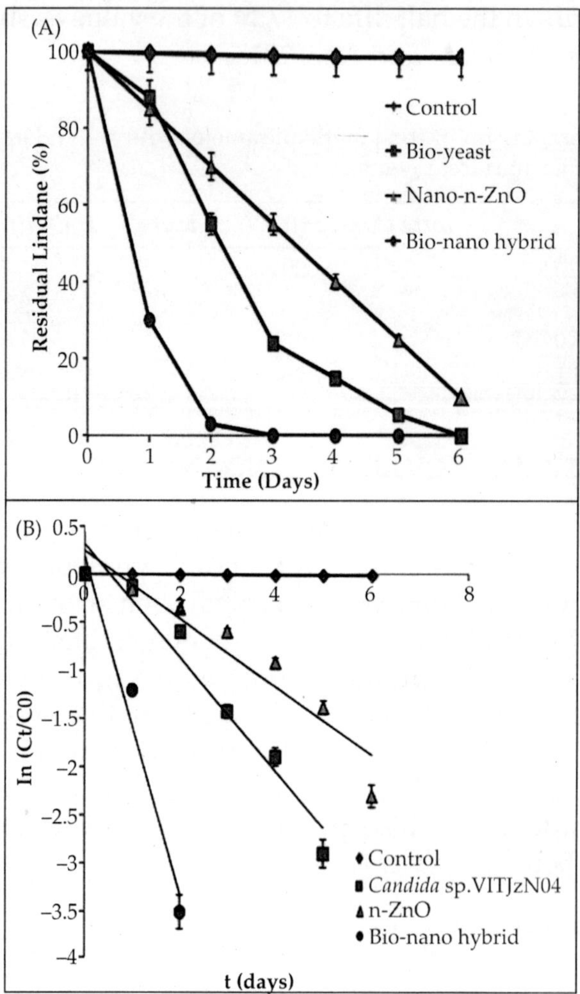

Fig. 7.9: Degradation of lindane in mineral medium. (A) Residual lindane percentage in the culture medium treated with various degradation agents (B) First–order kinetic plot of lindane degradation by various treatments.

of lindane on the pretreated biomass of VITJzN02 and VITJzN03 were fitted to various kinetic models *viz.* rst–order, second–order, intra–particle diffusion and Elovich models. Pseudo–first order model was found to be the best fitted model for lindane sorption onto the pretreated VITJzN02 and VITJzN03 biosorbents, indicating physical mode of adsorption process. Table 7.3 shows the reported works on lindane biosorption by various adsorbents. Maximum biosorption was noted in case of pretreated yeast *C. sorghi* VITJzN02 and *Rhodotorula* VITJzN03 (Salam and Das, 2013c).

Table 7.3: Reported works on lindane sorption using various adsorbents.

Sorbent	Q_{max} *(mg/g)*	*Reference(s)*
R. arrhizus	3.622 mg/l	Tsezos and Bell., 1989
Pretreated *E. coli*	4 mg/l	Ju *et al.*, 1997
Fly ash bagasse	2 μg/g	Gupta *et al.*, 2002
Granular activated carbon	487 ± 72 mg/g	Sotelo *et al.*, 2002
Pine bark	3.17 mg/g	Ratola *et al.*, 2003
Compost soil	48.42 ± 1.8 mg/g	Rama Krishna and Philip, 2008
Pretreated Powdered activated carbon	1.5 mg/l	Hassan *et al.*, 2009
Pretreated *C. sorghi* VITJzN02	100 mg/g	Salam and Das (2013c)
Pretreated *Rhodotorula* VITJzN03	88 mg/g	Salam and Das (2013c)

8

Remediation of Caffeine Using Yeasts

Caffeine is an alkaloid, found in natural products like coffee, tea, cocoa beans, guarana, cola nuts, and other plants. It is one of the main ingredients in a variety of beverages and caffeinated soft drinks and numerous food products. With the recent reemergence of medicinal herbs as a major player in the global dietary supplement market, new products–containing caffeine has been introduced. Of these, dry extracts of caffeine containing herbs and carbonated beverages, known as power or energy drinks, enriched with pure caffeine or caffeine extracts are becoming popular (Armenta *et al.*, 2005). It is widely used in pharmaceutical preparations as it enhances the effect of certain analgesics and antipyretic drugs. Caffeine is also used as a cardiac, neurological and respiratory stimulant and as a diuretic (Mazzafera, 2002). Because of its extensive use in food, beverages and medicines, caffeine has been detected in surface water, ground water and wastewater worldwide (Glassmeyer *et al.*, 2005).

Caffeine shows toxicity when fed in excess and is mutagenic *in vitro*. Excessive consumption of caffeine through beverages is associated with a number of health problems (Babu *et al.*, 2005). Intake of caffeine in large amounts over extended period of time can lead to a condition known as caffeinism. Reports are available on the toxic effects of caffeine on women, animals and plants (Pincheira *et al.*, 2003; Meyer *et al.*, 2004).

In recent years, caffeine has been considered as an emerging pollutant (Bernabeu *et al.*, 2012; Rivas *et al.*, 2012). The emerging contaminants may pose a potential hazard to humans and aquatic organisms (Bolong *et al.*, 2009). Presence of caffeine in soil affects soil fertility and it inhibits seed germination and growth of seedlings (Batish *et al.*, 2008). Caffeine is considered as a xenobiotic compound (Spiff and Uwakwe, 2003). Wastewaters containing caffeine are discharged to the surrounding water bodies and cause environmental pollution and health hazards. The ingestion of caffeine and its chlorinated byproducts (derived during chlorination of water during recycling) has severe adverse effects on physiological system (Dash and Gummadi, 2006a). Caffeine is detected in ground water, surface water and also in wastewater having high concentration (~10 g/L) (Buerge *et al.*, 2003; Weigel *et al.*, 2004; Glassmeyer *et al.*, 2005). To make the various water sources free from this xenobiotic, remediation of caffeine becomes a necessary step in the treatment of caffeine containing wastewaters (Dash and Gummadi, 2007a).

Though caffeine is considered to be toxic for many microorganisms, some microorganisms have the ability to grow in the presence of caffeine and have the capacity to utilize the same as a source of energy. The catabolic routes, however, are different in different groups of microbes reflecting the complexity in microbial caffeine degradation (Dash and Gummadi, 2006b). Caffeine degrading ability has been reported in various bacterial and fungal species isolated from the soil of coffee plantation areas or from coffee dumps. There are reports on bacteria and fungi, which have shown their potentiality for degrading caffeine. Reports are scanty regarding the potentiality of yeast as caffeine degrader. Among microorganisms, yeast cells offer several advantages. Yeasts are inexpensive, readily available sources of biomass. They not only grow rapidly like bacteria, but like filamentous fungi they also have the ability to resist unfavourable environments like extreme pH, temperature, depletion of nutrient sources, etc. (Yu and Wen, 2005).

CAFFEINE AND ITS PROPERTIES

Caffeine, an alkaloid of the methylxanthine family is a naturally occurring substance found in the leaves, seeds or fruits of over

63 plant species worldwide. The most commonly known sources of caffeine are coffee, cocoa beans, cola nuts and tea leaves (Heckman *et al.*, 2010). In its pure state, it is an intensely bitter white powder. Its chemical formula is $C_8H_{10}N_4O_2$ and its systematic name is 1, 3, 7–trimethylxanthine (Aurnaud, 1987). Its chemical structure is given in Fig. 8.1.

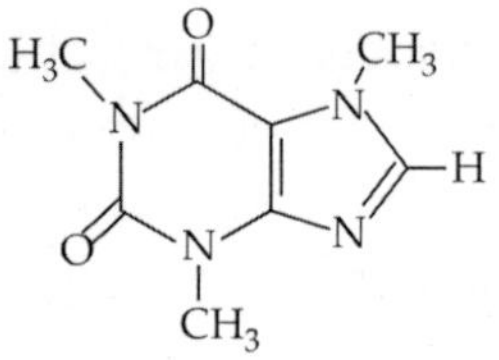

Fig 8.1: Caffeine (1,3,7–trimethylxanthine).

Caffeine was discovered by a German chemist, Friedrich Ferdinand Runge, in 1819. The structure of caffeine was given by Hermann Emil Fischer, who was also the first to achieve its total synthesis. Being readily available as a byproduct of decaffeination, caffeine is not usually synthesized. If desired, it may be synthesized from dimethylurea and malonic acid.

TOXICITY OF CAFFEINE

Environmentally, caffeine has been suggested as a chemical indicator of pollution for aquatic ecosystem since it is difficult to metabolize (Ogunseitan, 2002). Release of some caffeine–containing wastes, such as coffee pulp/husk from coffee producing factory, infused tea leaves, sewage from coffee industry and the factory of tea polyphenols preparation leads to serious ecological pollutions and becomes a big disposal problem (Pandey *et al.*, 2000). Discharging of unconsumed tea to the stream and sea is also a reason for the pollution of caffeine in the environment (Yalova *et al.*, 2003).

Aquatic Toxicity

Caffeine can persist in aquatic environments and has the potential to biomagnify through the food chain (Seiler *et al.*, 2005). Thus, even a small amount of dissolved caffeine in aquatic environment can be concentrated over time. Bantle *et al.* (1994) found that caffeine concentrations in water were high enough to affect *Xenopus laevis* egg development when exposed for 96 hours (LC_{50} = 0.22 to 0.37 mg/mL). Based on the report of International Uniform Chemical Information Database (2004), it was found that eggs of X. *laevis* were exposed to caffeine for 120–hour, LC_{50} values were between 0.13 and 0.19 mg./mL. Effect of caffeine on algal growth was studied by Pollack *et al.* (2009). The minimum inhibitory concentration (MIC) of caffeine was 30 mg/L for *Symbiodinium goreaui* and *Symbiodinium* sp. isolated from *Pseudoterogorgia bipinnata,* 50 mg/L for *Symbiodinium microadriaticum,* and 75 mg/L for *Symbiodinium* sp. isolated from *Aiptasia pallada.* Agrawal (1985) found that treatment of vegetative cells of S*tigeoclonium pascheri* with caffeine delayed the initiation of sporulation and decreased the percentage of sporulation. Moore *et al.* (2008) reported that the freshwater species *viz., Ceriodaphnia dubia, Pimephales promelas* and *Chironomus dilutus* were sensitive to caffeine exposure. *C. dubia* were found to be more sensitive to caffeine exposure (LC_{50} = 60 mg/L) than either *P. promelas* (LC_{50} = 100 mg/L) or *C. dilutus* (LC_{50} = 1,230 mg/L).

Microbial toxicity

The toxic effects of caffeine on microorganisms have been reported by many workers. Caffeine inhibited incorporation of thymidine into DNA in *Escherichia coli.* It induced filamentous growth of the same strains and inhibited the enzyme, thymidine kinase (Sandlie and Kleppe, 1980). Caffeine in small concentrations caused stimulation of the incorporation of dNMP into DNA. At large concentrations, however, it inhibited the DNA synthesis reaction (Sandlie and Kleppe 1982). Caffeine inhibited DNA repair in bacteria as a method to introduce mutants. Caffeine at 0.1% concentration also inhibited protein synthesis in bacteria (Grigg, 1972; Middelhoven and Lommen, 1984). Ramanaviciene *et al.* (2003) investigated the direct effect of caffeine on gram–negative bacteria *viz., Escherichia coli* and *Pseudomonas fluorescens*. They showed that the growth of *E. coli* DH5α and *P. fluorescens* 5443 was affected by caffeine. At caffeine concentration (1%) in the culture medium, the growth of both bacteria was inhibited. The optical density which represented the growth of *E. coli* and *P. fluorescens* cultures decreased 2.7 and 3.1 times, respectively, in comparison with control cultures grown in the medium without caffeine. The toxic effect of caffeine on *E. coli* was observed at 0.1% caffeine concentration. Below this concentration, caffeine showed no effect on the bacterial growth. Significant effect of caffeine on *P. fluorescens* colony growth was observed at 1% concentration of caffeine. Ibrahim *et al.* (2006) reported that *E. coli* could grow in broth containing 0.25% caffeine. Bacterial growth was significantly decreased compared to control. Increasing concentrations of caffeine caused inhibition of growth for all *E. coli* strains tested in this study. With the use of 0.75% caffeine, bacterial population was reduced by 1.4 log CFU/ml.

Plant Toxicity

Caffeine shows toxicity against plant species, but the mechanism of action has not yet been elucidated. A study was conducted to determine the effect of caffeine on the rooting of hypocotyl cuttings of mung bean (*Phaseolus aureus*) and the associated biochemical changes (Batish *et al.*, 2008). At lower concentrations (1,000 µM) of caffeine, though rooting potential was not affected, there was a significant decrease in the number of root and root length. At 1,000 µM caffeine, 68% decrease in the number of roots/primordia per cutting was noted and root length was decreased by over 80%. No root formation occurred at 2,000 µM caffeine concentration. Further investigations on biochemical processes linked to root formation revealed that caffeine significantly affected protein content, activities of proteases, polyphenol oxidases (PPO) and total endogenous phenolic (EP) content in the mung bean hypocotyls. Ashihara and Crozier (2001) described caffeine as a less explored natural plant product. Effect of caffeine on growth of rice seedlings was investigated over a period of 6 days. Concentration in the range of 0.5–10 mM caffeine inhibited the growth of seedling. Shoot and root elongation was inhibited by 50% and 80% with 2.5 mM caffeine (Smyth, 1992). Rizvi *et al.* (1987) observed that caffeine reduced the activity of amylases in the germinating seedlings of *Amaranthus spinosus* and gibberellic acid could not counteract the inhibitory effect. Caffeine inhibited the growth of rice and maize seedlings and the effect was found to be more severe on roots than on shoots (Smyth, 1992; Anaya *et al.*, 2002). It also inhibited mitosis and cell plate formation in the root tips of *Coffea arabica* (Friedman and Waller, 1983a).

DELETERIOUS EFFECTS OF CAFFEINE ON HUMAN HEALTH

Caffeine enters the human system in the form of beverages like tea, coffee and caffeinated soft drinks and numerous food products like chocolates and desserts with global consumption ranging from 80 to 400 mg caffeine per person per day (Gokulakrishnan *et al.*, 2005). Intake of caffeine in large amounts over extended period of time can lead to a condition known as caffeinism which causes unpleasant physical and mental conditions including nervousness, irritability, anxiety, tremulousness and muscle twitching (hyperreflexia), insomnia, headaches, respiratory alkalosis, and heart palpitations. Caffeine also increases the production of stomach acid and hence high usage over a period of time can lead to peptic ulcers, erosive esophagitis, and gastroesophageal reflux disease (Leson *et al.*, 1988; Iancu *et al.*, 2007). Caffeine consumption during pregnancy increases the risk of spontaneous abortion and affects foetal growth (Fernandez *et al.*, 1998). In addition, moderate to high caffeine intake (>150 mg caffeine/day) imbalances the bone mineral density causing osteoporosis particularly in women (Cooper *et al.*, 1992).

CONVENTIONAL METHODS FOR CAFFEINE REMOVAL

To date, four major approaches for reducing caffeine content from caffeine–containing products are conventional breeding, physicochemical methods, genetic engineering and microbial degradation, whilst approach for wastes decaffeination is rarely concerned. Decaffeination through conventional breeding is difficult to achieve because of non–availability of the low–caffeine containing germplasms (Fang *et al.*, 2011). Genetically–modified (GM) plants with low caffeine can be directly obtained either by down regulating the caffeine synthesis pathway or by up–regulating the caffeine degradation pathway (Yadav and Ahuja, 2007). However, this approach cannot be applied recently because of their instability and ecological risk of GM plants, even though successful study in laboratory scale was reported (Kato *et al.*, 2000). Decaffeination with special reagents extraction (Sun *et al.*, 2010) and chromatograph separation is usually adopted in practices (Ye *et al.*, 2009; Lu *et al.*, 2010; Dong *et al.*, 2011), although these techniques need to be optimized furthermore since their lack of security or/and yield. Comparatively, decaffeination through microbial degradation is more beneficial and attractive than other methods since it can be conducted safely and at a low budgetary requirement (Ramarethinam and Rajalakshmi, 2004; Gummadi and Santhosh, 2006).

PHYSICAL METHODS OF CAFFEINE REMOVAL

Water Extraction

Caffeine can be extracted using water as polar solvent. In the last few years, sub critical water (also called as high–temperature water, super heated water, pressurized hot water or hot liquid water by different investigators) has received increasing attention as an alternative extraction fluid (Li *et al.*, 2000). In water decaffeination method, caffeine is first extracted in water and then removed from water by solvent extraction process.

Charcoal and Carbon Filtering

Caffeine is removed from solvents by various processes including evaporation, charcoal addition, filtration and crystallization (Feldman and Katz, 1977). Charcoal and carbon filtering processes were developed as a direct challenge to solvent–based methods. Using only water, coffee elements are extracted from the beans, filtered through carbon or charcoal to remove the caffeine, and then the extract is replaced back to the bean. The patented Swiss Water Process is considered as a superior method for preserving flavor, because it throws away the first batch of beans and uses the decaffeinated coffee extract to wash and filter the next batch of beans, and so forth. Basically the difference is they are not using pure water to filter the beans, they are using "flavor charged" water that is already saturated with flavor ingredients, so only caffeine moves from the beans to the water. Thus, there is no need of re–soaking or re–infusing for bringing the removed flavor back into the coffee bean, because the flavor has not been removed.

CHEMICAL METHODS OF CAFFEINE REMOVAL

Extraction by Organic Solvents

In the solvent extraction method, caffeine is extracted by using organic solvents such as acetone, methanol, ethanol and acetonitrile as extraction solvents. Even though effective decaffeination can be achieved using the organic solvents, the use of the organic solvents is not appropriate because residual organic solvents have potential adverse effects on human health. Wang and Helliwell (2000) used solvent extraction to obtain caffeine–free green tea. Udayasankar *et al.* (1983) also reported the use of solvent extraction method for removal of caffeine. Caffeine sources were pre–treated with hot distilled water and caffeine was extracted with various solvents such as trichloroethylene, methylene chloride and similar chlorinated compounds. The optimal value of solvent to material ratio was reported as 1:4. This method is not widely used because of the environmental restrictions while discarding the solvents. Chloroform and methylene chloride are carcinogenic and have several human health concerns. Methylene chloride causes lesser hazard. Chlorinated hydrocarbon waste has significant environmental impacts and is costly to dispose. To overcome this problem, supercritical fluids are used as solvents.

Supercritical Fluid Extraction Using Carbon Dioxide

In this method, the caffeine source is placed in an extractor where supercritical carbon dioxide is let in and caffeine is removed from caffeine laden CO_2 under sub–critical conditions (Udayasankar *et al.*, 1986). Caffeine in solvent can be removed with the help of mesoporous silica and microporous silicalite membranes (Nwuha, 2000; Tan *et al.*, 2003). The advantage of utilizing supercritical carbon dioxide extraction ($SCCO_2$) have been well–documented (Roselius *et al.*, 1981; Roselius, Ludwig, Kurzhals, Hans-Albert, Hubert, Peter, 1981. Method for the selective extraction of caffeine from vegetable materials–US 4, 255, 458. Martinelli *et al.*, 1991). Extracts with solvent–free and non–toxic residues can be achieved. The viscosity of $SCCO_2$ is 5–15 times smaller than conventional organic or aqueous solvents. This leads to facile penetration of CO_2 into sample matrix and rapid mass transfer of desired extracts from the sample matrix to the $SCCO_2$ phase (Mohamed and Mansoori, 2002). In contrast to conventional organic or aqueous solvents, solvent power of $SCCO_2$ is very sensitive to small changes in experimental parameters such as pressure and temperature. Thus, selective extraction of desired compounds can be possible by simply adjusting the experimental parameters. In addition, the solvent power of $SCCO_2$ can be easily modified with an addition of entrainers or cosolvents. When the desired extracts are polar compounds, yield of the extracts can be substantially increased by adding small amount of polar solvents such as ethanol, acetone and water (Teberikler *et al.*, 2001). Park *et al.* (2007) reported the extraction of caffeine from green tea using $SCCO_2$/ethanol and $SCCO_2$/water. When $SCCO_2$/8.8 wt % water was used at 300 bar and 70°C, the caffeine extraction yield was 75.6%. When $SCCO_2$/ 7 wt % ethanol was used at 300 bar and 70°C, almost complete caffeine extraction yield was 97.3%.

Microwave Assisted Extraction Process

Caffeine and polyphenols from tea have been removed much more effectively by using microwave assisted extraction process (MAE) and solid phase extraction using molecularly imprinted polymer prepared with caffeine as a template (Theodoridis and Manesiotis, 2002; Pan *et al.*, 2003). Compared with the conventional extraction methods, the MAE procedure provided high extraction, high extraction selectivity, requiring short time, and less labour intensive.

Conventional decaffeination techniques like solvent extraction or use of supercritical carbon dioxide method can be expensive and toxic to the environment. These methods often results in the removal of aroma and flavour imparting substances and their precursors resulting in an unpalatable product. To compensate this, artificial flavours and colours are often supplemented to the decaffeinated product. Hence there is a strong need for alternative routes other than conventional extraction techniques for decaffeination process.

BIODEGRADATION OF CAFFEINE BY YEAST

The potential use of microorganisms and enzymes obtained from microbial system for developing biological decaffeination techniques offer a much attractive alternative to the present existing techniques (Gokulakrishnan *et al.*, 2005). Microbial biodegradation is an important biological method and promising decaffeination approach because of its low cost and high security (Fan *et al.*, 1999).

Five yeast species *viz. Trichosporon asahii, Candida tropicalis, Candida krusei, Candida inconspicua* and *Kloeckera japonica* isolated from caffeine contaminated samples were screened for their capability of caffeine degradation. Based on the degradation efficiency, isolate was selected and identified as *Trichosporon asahii* which showed 60% degradation of caffeine in 72 hours when caffeine was utilized as the sole carbon and nitrogen source. The influence of various factors such as pH, temperature, shaking speed, inoculum size, carbon source, nitrogen source and initial caffeine concentration on caffeine degradation were studied. The optimum growth conditions for caffeine degradation by *T. asahii* were found to be pH 6.5, temperature 28°C, shaking speed 120 rpm, inoculum size 4% (w/v) with initial caffeine concentration 2 gm/l and 100% degradation was achieved within 96 hours in the presence of sucrose (5 gm/l). The addition of external nitrogen sources *viz.* sodium nitrite and sodium nitrate decreased the caffeine degradation to 25% and 30% respectively. This was the first report on caffeine degradation using yeast cells (Lakshmi and Das, 2010).

Sixteen yeast strains were isolated from the cultivated tea soils collected from sites of northern Iran and evaluated for the caffeine tolerance by the agar dilution method (Ashengroph and Borchaluei, 2013). Based on the tolerance efficiency, morphological and biochemical characteristics as well as molecular studies strain TFS9 was selected and identified as *Saccharomyces cerevisiae* TFS9 (GenBank accession number KF414526). The time course of caffeine removal by growing cells of the strain TFS9 in the minimal salt medium containing caffeine as the sole source of carbon was estimated. The concentration of caffeine in the supernatant of the yeast culture medium decreased by 84.8% (from 3.5 g/l to 0.53 g/l) after 60 h of incubation by using of *S. cerevisiae* TFS9, without additional optimization process. Results of the experimental studies suggested that it was the first evidence on caffeine bio–degradation using yeast *S. cerevisiae.*

ENZYMATIC MECHANISM OF CAFFEINE DEGRADATION

Involvement of Cytochrome P–450 monoxygenase enzyme system in yeast *Trichosporon asahii* in caffeine liquid medium (CLM) was detected in the presence of inhibitor, 1–aminobenzotriazole (ABT) (Lakshmi and Das, 2011). The results showed that the enzyme activity was only present in yeast cells grown in caffeine containing medium and no activity was detected in the absence of caffeine confirming that caffeine is a good inducer of cytochrome P–450 enzymes in yeast isolate. Similar results were reported in case of other filamentous fungi (Cha *et al.*, 2001) and white rot fungi (Kanaly and Hur, 2006). The capacity of human Cytochrome P–450 monooxygenase to metabolize caffeine yielding trimethyl uric acids, paraxanthine and minor amounts of theobromine has been reported by Tanaka *et al.* (1993). Involvement of other enzymes *viz.* caffeine demethylase and xanthine oxidase activities were also observed in microsomal and cytosolic fractions of *Trichosporon asahii* (Lakshmi and Das, 2011). Analysis of GC data of caffeine showed retention time 11.75 min with molecular weight 194.1 (Fig. 8.2).

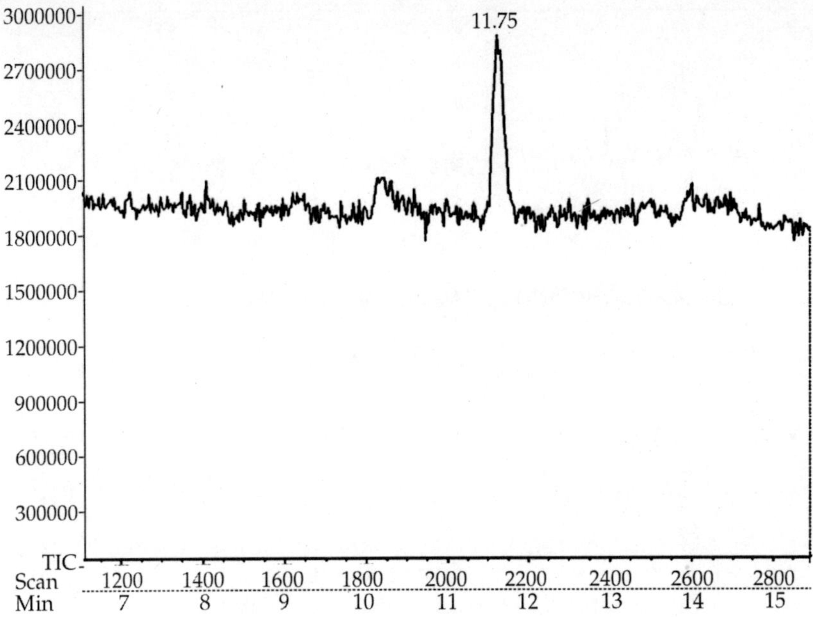

Fig 8.2: Gas chromatogram of caffeine.

Gas chromatogram of the degradation products obtained at 48 h showed the existence of four intermediates (Fig. 8.3a). The four intermediates with retention time 7.11, 8.29, 8.45 and 9.58 min were identified as xanthine with molecular weight 152.11 methylxanthine with molecular weight 166.13, uric acid with molecular weight 168.11 dimethylxanthine with molecular weight 180.64 respectively. Gas chromatogram of degradation products formed by *Trichosporon asahii* respectively did not show the existence of any compound which confirmed the complete degradation of caffeine (Fig. 8.4b).

Based on the results of enzyme activities and analysis of caffeine metabolites, the major degradation pathway of caffeine by *T. asahii* was proposed as shown in Fig. 8.4. According to the proposal, degradation of caffeine occurs *via* stepwise demethylation and oxidation processes. The sequence of products formed during caffeine degradation is as follows: 1,7–dimethylxanthine,1,3–dimethylxanthine, 1, 3 and 7–methylxanthines, xanthine and uric acid.

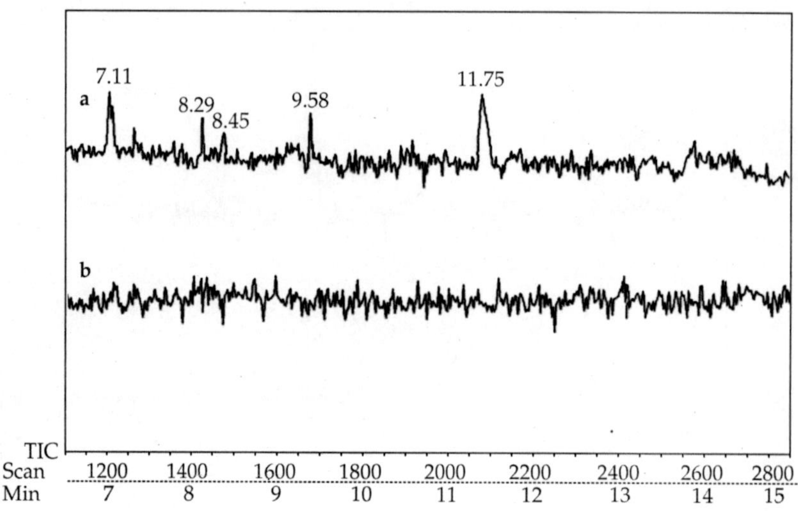

Fig. 8.3: Gas chromatogram of degradation products of caffeine formed by *Trichosporon asahii* at (a) 48 h and (b) 54 h. Medium: CLM, pH: 6.5, Temperature: 28°C (Lakshmi and Das, 2011).

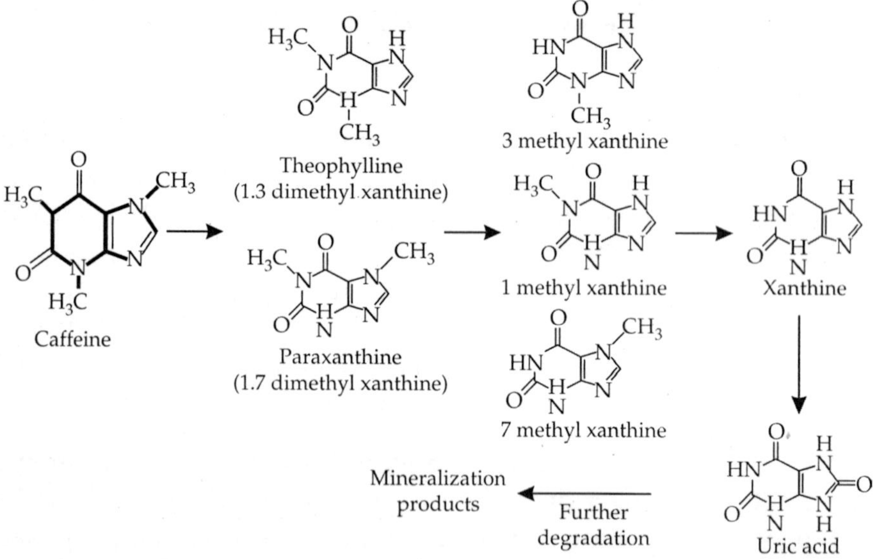

Fig. 8.4: Proposed biodegradation pathway of caffeine by *T. asahii* (Lakshmi and Das, 2011).

FTIR spectra of control caffeine (Fig. 8.5 a) showed the specific peaks in fingerprint region (3500–500 cm^{-1}). The peak at

3338.42 cm^{-1} supports for H–C–H stretching in methane group. The peak at 3100.74 cm^{-1} corresponds to aromatic –CH stretching vibrations. The peak at 2850.27 cm^{-1} indicates –CH stretch. The peak at 1700.75 cm^{-1} corresponds to C=0 stretch of ketones. The peaks at 1637.81 and 1483.83 cm^{-1} corresponds to C=C and C–C stretches respectively. The other peaks at 1547.00 cm^{-1} and below 1483.83 cm^{-1} represents C–N stretches. FTIR spectra of degradation products (Fig. 8.5b) showed different peaks at 3423.56 cm^{-1} for the presence of N–H stretching and 3111.16 cm^{-1} for NH_3^+ stretching. Absence of peaks at 3338.42 and 3100.74 cm^{-1} indicated the demethylation of caffeine. The FTIR spectra data supported the complete degradation of caffeine by *T. asahii.*

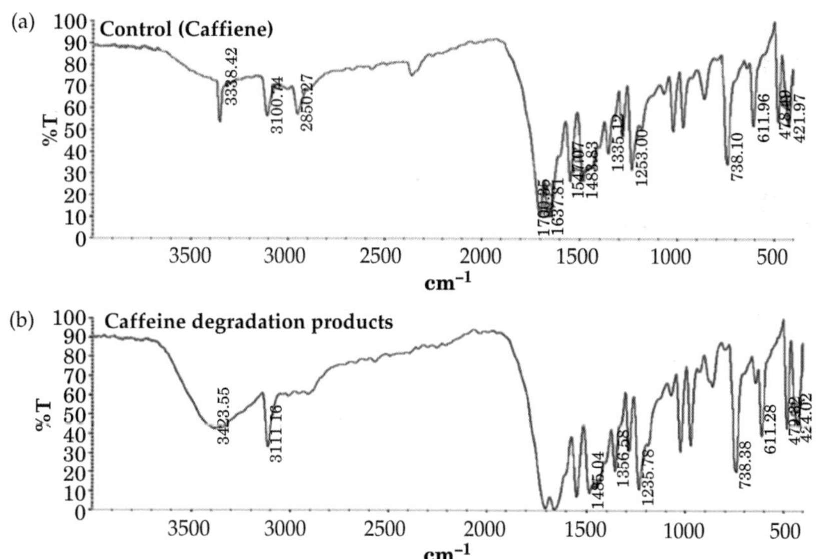

Fig. 8.5: FTIR spectral analysis of control, caffeine (a) and its degradation products (b) (Lakshmi and Das, 2011).

CAFFEINE DEGRADATION BY YEAST BIOFILM

Biofilms are structured microbial communities in which microbial cells irreversibly attach to a surface or interface and become embedded in a matrix of extracellular polymeric substances produced by these cells. Biofilms have been found to be suitable for the remediation of pollutants because of their high microbial biomass and ability to immobilize pollutants. Biofilm formation in microorganisms is closely linked with production of exopolysaccharides which act as glue, aiding in the attachment process and protect the underlying cells from fluctuations in the surrounding milieu (Saravanan *et al.*, 2008). Microbial exopolysacharides belong to a wide group of secreted polymers that can be adhered to the cell surface or released as extracellular slime in the surroundings of the cell (Knoshaug *et al.*, 2000). The biofilm forming ability of *Trichosporon asahii* was reported. The sequence of biofilm formation by the *T. asahii* in different time intervals was clearly seen in AFM (Fig. 8.6) and SEM images (Fig. 8.7). Thus it was evident that yeast species were capable of forming single–species biofilm (Lakshmi and Das, 2013a).

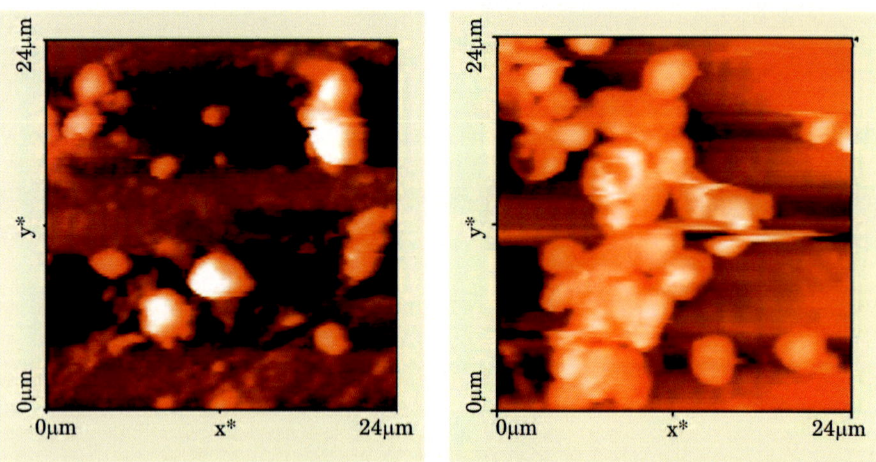

Fig. 8.6: AFM images showing sequential biofilm formation by *Trichosporon asahii* on PVC strips during (a) 6 h and (b) 12 h.

The exopolysaccharide (EPS) production by the yeast biofilm was clearly observed in the SEM images. Colorometric polysaccharide assay revealed that high level of EPS was achieved by *Trichosporon asahii* (Fig. 8.8) (Lakshmi and Das, 2013a). This EPS production might be the important mechanism of adhesion of yeast cells to an inert material to form biofilm. The extensive exopolysaccharide formation protects the microbial cells from environmental chemical toxicity (Chandran and Das, 2011).

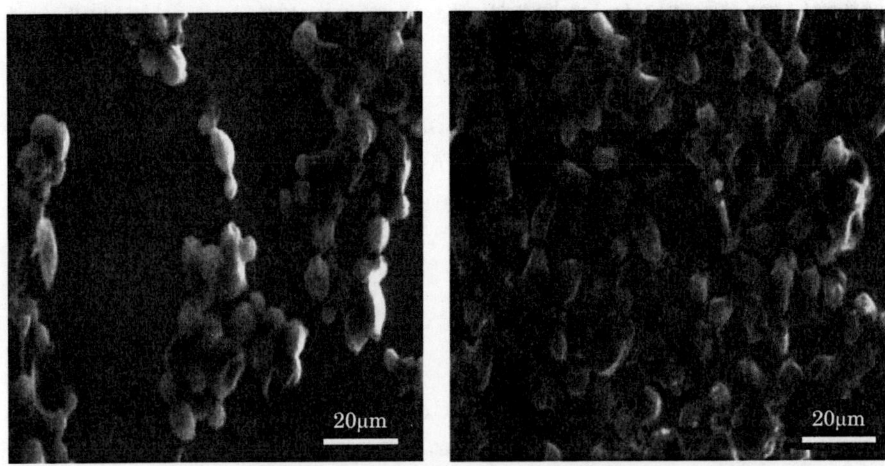

Fig. 8.7: SEM images (x 3500) showing sequential biofilm formation by *Trichosporon asahii* on PVC strips during (a) 18 h and (b) 24 h.

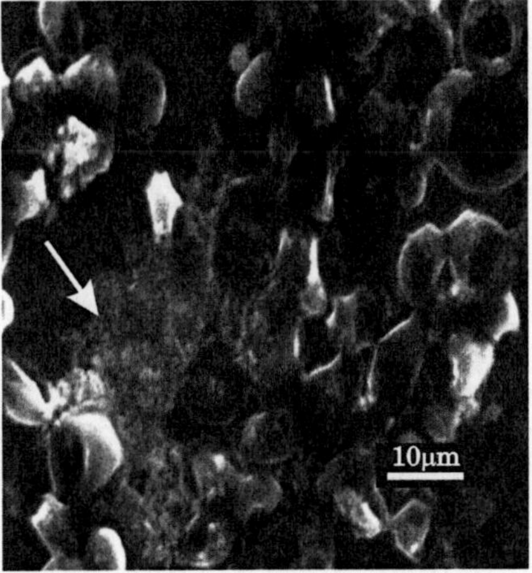

Fig. 8.8: SEM images (x 5000) showing production of exopolysaccharides by the biofilm formed by *Trichosporn asahii.*

CAFFEINE REMOVAL USING IMMOBILIZED YEAST

Immobilization technology is widely being used for the remediation of organic contaminants from the environment (Aksu and Bulbul, 1999). The efficacy of biodegradation is often improved when immobilized microbial cells are used. Immobilized microbial cells have several potential advantages over free cells for the bioremediation of anthropogenic wastes (Baskaran and Nemati, 2006). Moreover, the immobilized cells are less likely to be adversely affected by predators, toxins or parasites compared to free cells (Prabua and Thatheyus, 2007). Microbial cells immobilized in PVA–alginate hybrid matrix showed best performance in terms of degradation. This efficiency of PVA–alginate matrix is attributed to its better surface properties (Robledo–Ortíz *et al.*, 2011). Reports are scanty on degradation of caffeine by immobilized yeast cells.

Caffeine removal (through degradation) by yeast isolate, *Trichosporon asahii* immobilized on various conventional matrices (sodium alginate, carboxymethyl cellulose, chitosan, agar and agarose) was investigated using the method of entrapment (Lakshmi and Das, 2013a). Yeast cells immobilized in alginate showed more growth than those immobilized in other substrates. This proved that alginate provided suitable support for the growth of yeast species in caffeine containing media (Fig. 8.9). Alginate immobilized yeast *T. asahii* showed higher percentage of caffeine degradation (75%), compared to the free cells (70%). This was the first report showing more caffeine degradation capacity of immobilized yeast *T. asahii* compared to free cells.

Lakshmi *et al.* (2013b) reported that the yeast species *T. asahii* was immobilized in single and hybrid matrices such as Na–alginate, polyvinyl alcohol (PVA)–alginate and glycerol–alginate and used for caffeine degradation. Degradation of caffeine and growth of *T. asahii* was optimum when 8 g/L sucrose and 10 g/L caffeine was present in the medium. PVA–alginate immobilized cells showed the maximum caffeine degradation compared to free and other immobilized cells. Bead diameter showed profound effect on biodegradation rate, diffusivity, effectiveness factor and Thiele modulus. Bead parameters (bead diameter, Na–alginate concentration, PVA concentration and cell loading) of PVA –alginate hybrid bead were optimized using 2D factorial design with a high model significance ($p<0.0001$). Maximum caffeine degradation was obtained at a combination of bead diameter 0.27 cm, Na–alginate concentration 10 g/L, PVA concentration 50 g/L and cell loading 4 g/L (wet weight). In column mode, 98.9% caffeine degradation was obtained at optimized flow rate of 1 ml/min and bed height of 12 cm using PVA–alginate immobilized *T. asahii* cells. Successful attempt on caffeine removal from industrial wastewater was made using the same hybrid beads.

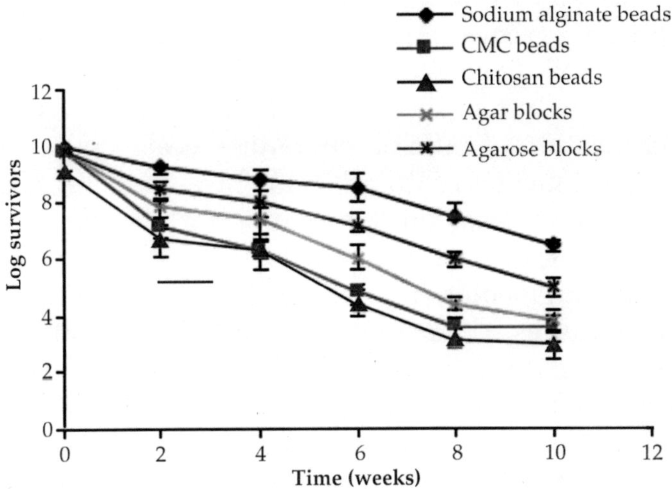

Fig. 8.9: Time–course of changes in the viability of *T. asahii* immobilized in various matrices (T= 28°C, pH = 6.5).

CAFFEINE REMEDIATION FROM SOIL USING YEAST

Biodegradation of caffeine in agricultural soil using liquid municipal biosolids as bulking agent has been reported (Topp *et al.*, 2006). This bulking agent enhanced the aeration and microbial activity, and thereby increased the biodegradation rate of contaminants. Bioaugmentation and biostimulation can accelerate the rate of biodegradation in soil (Vasudevan and Rajaram, 2001). In many cases, bioaugmentation was reported as a feasible strategy for contaminant removal and site remediation (Ruberto *et al.*, 2003; Jacques *et al.*, 2008). Addition of enriched indigenous microorganisms or exogenous microorganisms can degrade the toxic contaminants and offer higher tolerance to toxicity (Ellis *et al.*, 2000; Barathi and Vasudevan, 2003).

Immobilised yeast for the remediation of caffeine in soil under laboratory condition was tested (Lakshmi and Das, 2012). Biostimulation with inorganic nutrients and bioaugmentation with caffeine utilizing single and mixed yeast cultures *viz.* Mixed culture 1, Mixed culture 2, Mixed culture 3 and Mixed culture 4 were employed as remedial options for the removal of caffeine from contaminated soil. To promote caffeine removal, biowaste materials (wheat bran, sawdust, peanut hull powder) were used as biocarriers for immobilization of caffeine degrading yeast species. Fig. 8.10 shows the SEM images of the adhesion of *T. asahii* (a), *C. tropicalis* (b) and (c) *C. inconspicua* onto the surface of sawdust. Laboratory biopiles was constructed to compare the treatment bioaugmentation alone with bioaugmentation combined biostimulation using single and mixed cultures of immobilized yeast. Maximum removal of caffeine was found to be 98.3% after 24 days in the biostimulated soil which was bioaugmented with sawdust immobilized yeast mixed culture No. 4 consisting of three yeast species *viz. Trichosporon asahii, Candida tropicalis* and *Candida inconspicua*. Dehydrogenase activity in the soil was remarkably enhanced to 639 µg TPF g^{-1} soil and microbial numbers were also increased for the soil treated under the same conditions described above. Phytotoxicity assay confirmed the reduction of caffeine toxicity in the contaminated soil after treatment. Thus, sawdust immobilized mixed yeast culture No. 4 could serve as potential tool for the remediation of caffeine from contaminated soil.

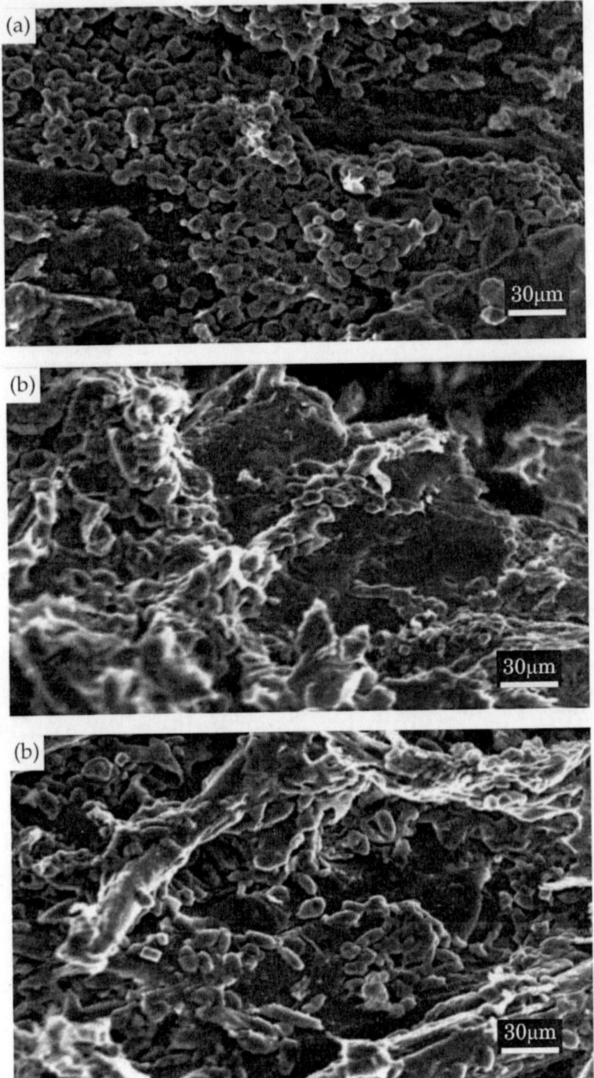

Fig. 8.10: SEM images (x 1500) showing the adhesion of *T. asahii* (a), *C. tropicalis* (b) and *C. inconspicua* (c) onto the surface of sawdust following 48 h incubation and washing with 10 mM phosphate buffer (pH 7.0) (Lakshmi and Das, 2012).

9

Degradation and Removal of Antibiotics by Yeasts

Pharmaceuticals are indispensible for the maintenance of public health and quality of life.

Rapid advances in drug therapies to meet health challenges and their timely availability are essential for a healthy society (Caracciolo *et al.*, 2015). The presence of pharmaceutical compounds, especially antibiotics, in the ecosystem has been known for almost 30 years. However, their presence became an emerging concern only in mid–1990s (Lissemore *et al.*, 2006). Residues of human and veterinary antibiotics were detected in a multiplicity of matrices (Mompelat *et al.*, 2009). The introduction of these compounds into the environment through anthropogenic sources can constitute a potential risk for aquatic and terrestrial organisms. Although present at vestigial levels, antibiotics may cause resistance in bacterial populations making them ineffective in the treatment of several diseases in near future (Martínez, 2009). Recently, the widespread detection of pharmaceuticals in terrestrial and aquatic systems has endangered significant scientific and regulatory concern (Cardoso *et al.*, 2014). The characteristics of antibiotics include polymorphism, their introduction into the environment after human metabolism, their chemically complex structure, and the fact that they can be ionized and have multiple ionization sites spread throughout the molecule (Cunningham, 2008). Relevant processes regarding removal of antibiotics in the environment include sorption to soils and sediments, complexation with metals and organics, chemical oxidation, photolysis, volatilization, and biodegradation (Babic *et al.*, 2007).

ANTIBIOTICS

The first antibiotics were of natural origin, *e.g.*, penicillins produced by fungi in the genus *Penicillium*, or streptomycin from bacteria of the genus *Streptomyces*. Currently, antibiotics are obtained by chemical synthesis, such as the sulfa drugs (*e.g.*, sulfamethoxazole), or by chemical modification of compounds of natural origin. Many antibiotics are relatively small molecules with a molecular weight of less than 1000 Da. The classical definition of an antibiotic is a compound produced by a microorganism which inhibits the growth of another microorganism. The term antibiotic is now refers to substances with antibacterial, anti–fungal, or anti–parasitical activity. The most important classes and groups of antibiotic compounds are shown in Table 9.1 Antibiotics can be grouped by either their chemical structure or mechanism of action. They are a diverse group of chemicals that can be divided into different sub–groups such as β–lactams, quinolones, tetracyclines, macrolides, sulphonamides and others. They are often complex molecules which may possess different functionalities within the same molecule. Therefore, under different pH conditions, antibiotics can be neutral, cationic, anionic, or zwitterionic. Because of the different functionalities within a single molecule, their physico–chemical and biological properties (Cunningham, 2008), sorption behaviour, photo reactivity, antibiotic activity and toxicity may change with pH. In some Asian countries, concentrations of antibiotics and its compounds reached up to several mg/L in effluents. In developed countries, a manufacturing plant can also make a significant contribution to total antibiotic concentration in the influent of a sewage treatment plant (STP).

Table 9.1: Important classes and groups of antibiotic compounds.

Class	*Group*	*Sub group*	*Example*
β–lactams	Penicillins	Acylaminopenicillins	Piperacillin
		Aminopenicillins	Amoxicillin
		Benzyl–penicillins	Phenoxypenicillin
		Carboxypenicillins	Carbenicillin
		Isoxazolylpenicillins	Oxacillin
	Cephalosporins	Cefalexin group	Cefprozil
		Cefotaxime group	Cefotaxime
		Cefuroxime group	Cefuroxime
		Cefazolin group	Cefazolin
Glycopeptides		–	Vancomycin
Macrolides		–	Erythro–mycin A
Sulfonamides		–	Sulfamethoxazole
Tetracyclines		–	Doxycycline
Quinolones		–	Ciprofloxacin

SOURCES

Antibiotics are used extensively in human and veterinary medicine, as well as in aquaculture, for the purpose of preventing (prophylaxis) or treating microbial infections. Several hundred different antibiotic and antimycotic substances are used in human and veterinary medicine (Kümmerer and Henninger, 2003). Therefore, the use of antibiotics was widespread (annual consumption of 100000–200000 tons) and consequently, the possibility of water contamination with such compounds increased (Xu *et al.*, 2007). Antibiotics can be more or less extensively metabolized by humans and animals. After administration, antibiotics for human use or their metabolites are excreted into the effluent and reach the sewage treatment plant (STP). The non–metabolized fraction is excreted as a still–active compound. Residues of antibiotics have been detected in surface waters (Kim and Carlson, 2006; Xu *et al.*, 2007; Watkinson *et al.*, 2009), ground waters (Batt *et al.*, 2006; Xu *et al.*, 2007), sea waters (Minh *et al.*, 2009), drinking water (Yiruhan *et al.*, 2010), WWTPs effluents (Minh *et al.*, 2009) and hospital wastewaters (Watkinson *et al.*, 2009). Antibiotics have also been detected in terrestrial matrices and biosolids (Feitosa–

Felizzola and Chiron, 2009). Usually, antibiotics are detected in the high concentration (mg/L) in hospital effluents whereas lower concentration (µg/L) range in municipal wastewater and ng/L in surface, sea and groundwater. Moreover, sediments from agriculture–influenced rivers contain higher antibiotic concentrations than other sediments from rivers located far from agricultural areas. This indicates the possibility of run–off contamination from farmland (Kümmerer, 2009). Antibiotic residues have also been detected in soil due to application of manure or sludge as fertilizers to the soil. Presence of residues have also been reported in vegetables and cereals such as carrots, lettuces, green onions, cabbages, cucumbers and corn (Dolliver *et al.*, 2007; Shenker *et al.*, 2011).

The accumulation and persistence of antibiotics in the environment produce harmful effects, either in aquatic or terrestrial ecosystems, even in the low concentrations levels, in which they are detected (ng/L to mg/L for water matrices and low medium mg/kg for sediments).

REMEDIATION METHODOLOGIES

Being persistent and recalcitrant to biodegradation, most of the antibiotics are accumulating to the environment and causing pollution. The presence and fate of antibiotics in various environmental matrices have received a great deal of attention by the scientific community. Filtration and coagulation, flocculation, sedimentation are the conventional technologies mostly used for the treatment of wastewater containing antibiotics (Vieno *et al.*, 2007; Arikan, 2008). For example, toxic effects of many pollutants to the microorganisms and recalcitrant nature of pollutants prevent the application of activated sludge technology which is widely used in industrial effluent treatments (Britto and Rangel, 2008). Filtration is a process for the removal of the solids, specially suspended solids by passing the wastewater through a granular media (sand, coal, diatomaceous earth, granular activated carbon). This process does not degrade the contaminant, but concentrates the particles in the solid phase, generating a new waste. The processes like coagulation/

flocculation/sedimentation employ chemicals like lime, alum, iron salts and polymers to enhance the solids sedimentation. But these techniques require a subsequent treatment to remove the pollutants (in a coagulated form) from the effluents (Eckenfelder, 2007). But all these technologies were unsuccessful for the removal of these compounds due to some drawbacks and other technologies were proposed.

Other technologies mainly include oxidation processes which involve chlorination and advanced oxidation processes (AOP) along with adsorption, membrane processes and combined methods. Examples of AOPs include ozonation, Fenton, photo–Fenton, photolysis, semiconductor photocatalysis and electrochemical processes. Ozonation, Fenton's oxidation and semiconductor photolysis are the most tested methodologies. They were also considered as the most applied processes to the β–lactam class (the mainly prescribed antibiotics). Ozonation is effective in antibiotics removal and has the advantage of being applied to fluctuating flow rates and compositions. But the high cost of equipment and the energy required to supply the process constitutes the major drawbacks of the process (Dantas *et al.*, 2008; Li *et al.*, 2008). Although Fenton process produced good results in many cases (degradation efficiency above 53%, COD removal >44%, TOC removal >20% and a slightly increase in biodegradability), the photo–Fenton seems to be more efficient (degradation efficiency above 74%, COD removal >56%, TOC removal >50%) (Homem and Santos, 2011). The membrane technologies *viz.* reverse osmosis, nano and ultrafiltration are increasingly used as separation processes. However, this technology does not enable the removal or degradation of the contaminant, but only its transference to a new phase (the membrane), where it is present in a more concentrated form. Adsorption is another process that has been reported as an alternative to oxidation techniques, though not widely applied to the more prescribed antibiotics. This technique was reported as very efficient (removals above 80%). However, it has the disadvantage of producing a new residue. Activated carbon, a high cost adsorbent material was used as adsorbent in most of the cases which is not a cost effective approach.

Table 9.2 shows the treatment methods applied for the remediation of antibiotics based on the pollutant concentration in the effluent and the cost of the process.

Table 9.2: Treatment processes adopted for remediation of antibiotics.

Antibiotics	*Concentration*	*Treatment*	*Reference(s)*
Ampicillin	20 mg/L	Photo–Fenton	Rozas *et al.*, 2010
Amoxicillin	1–100 mg/L	Semiconductor photo-catalysis	Klauson *et al.*, 2010
Enrofloxacin	158–790 mg/L	Anodic oxidation with Electrogenerated H_2O_2 Electro–Fenton Photo-electron–Fenton Solar photoelectron–Fenton	Guinea *et al.*, 2010
Dimetridazole Metronidazole Ronidazole Tinidazole	150 mg/L	Adsorption on activated carbons	Méndez–Diaz *et al.*, 2010
Amoxicillin	300 mg/L	Adsorption on activated carbon and bentonyte	Putra *et al.*, 2009
Amoxicillin	500 mg/L	Photo– Fenton	Elmolla and Chaudhuri 2009
Amoxicillin	42 mg/L	Photo–Fenton	Trovó *et al.*, 2008
Flumequine	19.1– 95.7 mM	Semiconductor Photo-catalysis	Palominos *et al.*, 2008
Penicillin G	600 mg/L	Ozonation	Arslan–Alaton and Caglayan 2006
Amoxicillin	5.0×10^{-4M}	Ozonation	Andreozzi *et al.*, 2005
Penicillin	830 mg/L	Ozonation	Arslan Alaton and Dogruel 2004
Penicillin	720 mg/L	Ozonation	Cokgor *et al.*, 2004

BIODEGRADATION OF ANTIBIOTICS BY YEASTS

The physico–chemical methods used for the treatment of pharmaceutical wastewaters which are of limited applicability because of limitations such as inefficiency of remediating high strength wastewater, high operating cost, huge labour requirement, high equipment cost, intervention of toxic by–products etc. (Homem and Santos, 2011). Thus, one way to solve these problems to use biodegradation as remediation technology involving microorganisms which is attracting increasing attention now–a days as a less expensive and more environmentally friendly alternative to the conventional treatment methods (Okoh, 2006). Most antibiotics tested to date have not been biodegradable under aerobic conditions. Gartiser *et al.* (2007) reported that ampicillin, doxycycline, oxytetracycline, and thiamphenicol were significantly degraded, while josamycin remained at initial levels. Tylosin was biodegraded (Hu and Coats, 2007). Out of 16 antibiotics tested, only benzyl penicillin (penicillin G) was completely mineralized in a combination test. Trials simulating sewage treatment with radiolabeled compounds revealed that approximately 25% of benzyl penicillin was mineralized within 21 d, whereas ceftriaxone and trimethoprim were not mineralized at all (Junker *et al.*, 2006).

There are reports on the use of bacteria *viz. Pseudomonas putida* and *Pseudomonas fluorescens* (Krishnan *et al.*, 2012), *Bacillus* and *Bacteroides* (Wagner *et al.*, 2011) for degradation of cephalosporin derivatives. However, reports are scanty on yeast as remediation agent for removal of antibiotics.

Four yeast species *viz. Pseudozyma sp.* SMN01, *Ustilago sp.* SMN02, *Ustilago sparsa SMN03* and *Candida blankii SMN04* were isolated from pharmaceutical wastewater and screened for their ability to degrade third generation cephalosporin derivatives *viz.* cefdinir, cefotaxime and cefoperazone (Selvi and Das, 2014 a,b). Maximum cefdinir degradation efficiency was noted as 72% at 150 mg/L for *Pseudozyma sp.* SMN01, 78% at 200 mg/L for *Ustilago sp.*SMN02, 81% at 200 mg/L for *Ustilago sparsa* SMN03 and 84% at 250 mg/L for *Candida blankii* SMN04 respectively. In a study by Hamrapurkar *et al.* (2011), cefdinir degradation of 48.83% by base hydrolysis method was reported. There was a report on the cephalosporin degradation of concentration 175 mg/L which was removed upto 81% after 90 days of operation in a reactor (Sundararaman and Saravanane, 2010) In another study, maximum biodegradation efficiency of cephalosporin derivatives using *Pseudomonas sp.* were reported as 64% (Krishnan *et al.*, 2012).

The study made by Selvi and Das (2014a) was the first report on yeasts showing the best cefdinir degrading potentiality at higher concentration. Therefore, it was noteworthy that the yeast isolates could tolerate 150–250 mg/L of cefdinir and degrade 72–84% at the end of 6 days which was found to be quite high when compared with other biological and physico–chemical methods of degradation of cephalosporin derivatives reported so far.

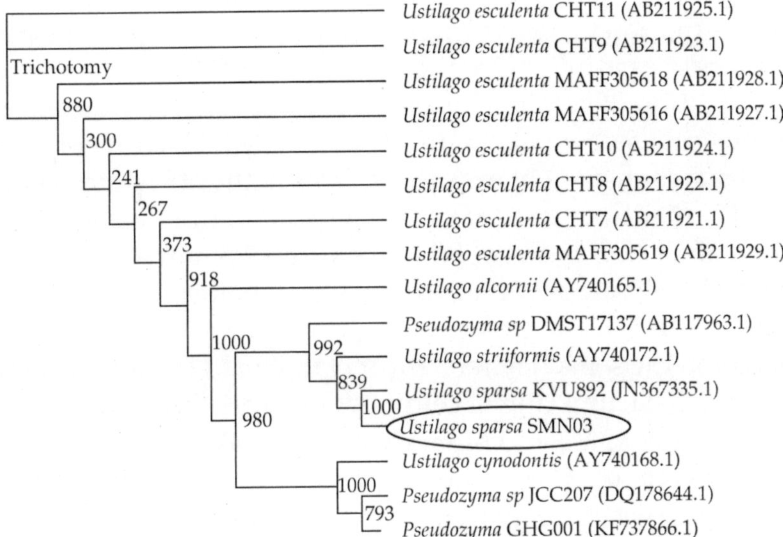

Fig. 9.1: Phylogenetic analysis of strain SMN03 and related species. Phylogenetic tree of partial and complete gene sequences of 18S rRNA, ITS–1, 5.8S rRNA and 28S rRNA. The tree was tested by bootstrap taking 1000 samples (Selvi *et al.*, 2014).

Selvi *et al.* (2014) reported the biodegradation of cefdinir by a novel yeast strain isolated from pharmaceutical wastewater. Molecular characterization suggested that the yeast belonged to the genus *Ustilago* and was named as *Ustilago sparsa* SMN03 (Fig. 9.1). It was noteworthy that the isolated yeast strain *Ustilago sparsa* SMN03 could tolerate 200 mg/L of cefdinir and degrade 81% at the end of 6 days which was found to be quite comparable to other biological and physico–chemical methods of degradation of cephalosporin derivatives reported so far. This observation was extremely important from the perspective of environmental pollution control.

The LC–MS spectra of cefdinir and its metabolites formed during degradation by *Ustilago sparsa* SMN03 at different time intervals *viz,* 0,2,4,6 and 7th day are shown in Fig. 9.2 a–e.

In Fig. 9.2 a, the day 0 sample showed a high intense peak of 1;0 retention time (RT) 1.6 min, corresponding to the m/z value of 396.1 in the positive ion mode which confirmed the presence of the parent compound, cefdinir. On the day 2 (Fig.9.2b), the onset of biodegradation of cefdinir by *Ustilago sparsa* SMN03 was observed by presence of a new peak with RT 4.1 min, which corresponded to the first metabolite M1 of m/z value 364.1 in the MS spectra. On day 4 (Fig. 9.2 c), the occurrence of parent peak with reduced intensity and two other new peaks of RT 1.9 min and 2.3 min were observed. These peaks corresponded to two new metabolites M2 (m/z=326.1) and M3 (m/z=338.4) respectively. The disappearance of M1 was observed on the same day too. The metabolite M2 structure showed the opened β–lactum ring of cefdinir. On day 6, the reduction in the peak intensities of M2 and M3 were observed along with the peak of the parent compound with no new compounds formed (Fig. 9.2d). But on day 7 (Fig. 9.2e), two other metabolites M4 (RT 1.2 min) and M5 (RT 0.8 min) with m/z value 170.2 and 99.4 were observed along with the parent compound cefdinir which was found to degrade 81 % as calculated from the liquid chromatogram. Based on the characterization of metabolites, a hypothetical biodegradative pathway of cefdinir degradation by yeast was proposed (Fig. 9.3).

FT–IR analysis showed that most of the major functional groups such as carboxyl group, hydroxyl group, lactam group and carbonyl group were removed or replaced showing parent compound degraded to simpler intermediates for utilization as carbon and energy source by the yeast isolate (Fig. 9.4). It was concluded that the novel yeast strain *Ustilago sparsa* SMN03, isolated from

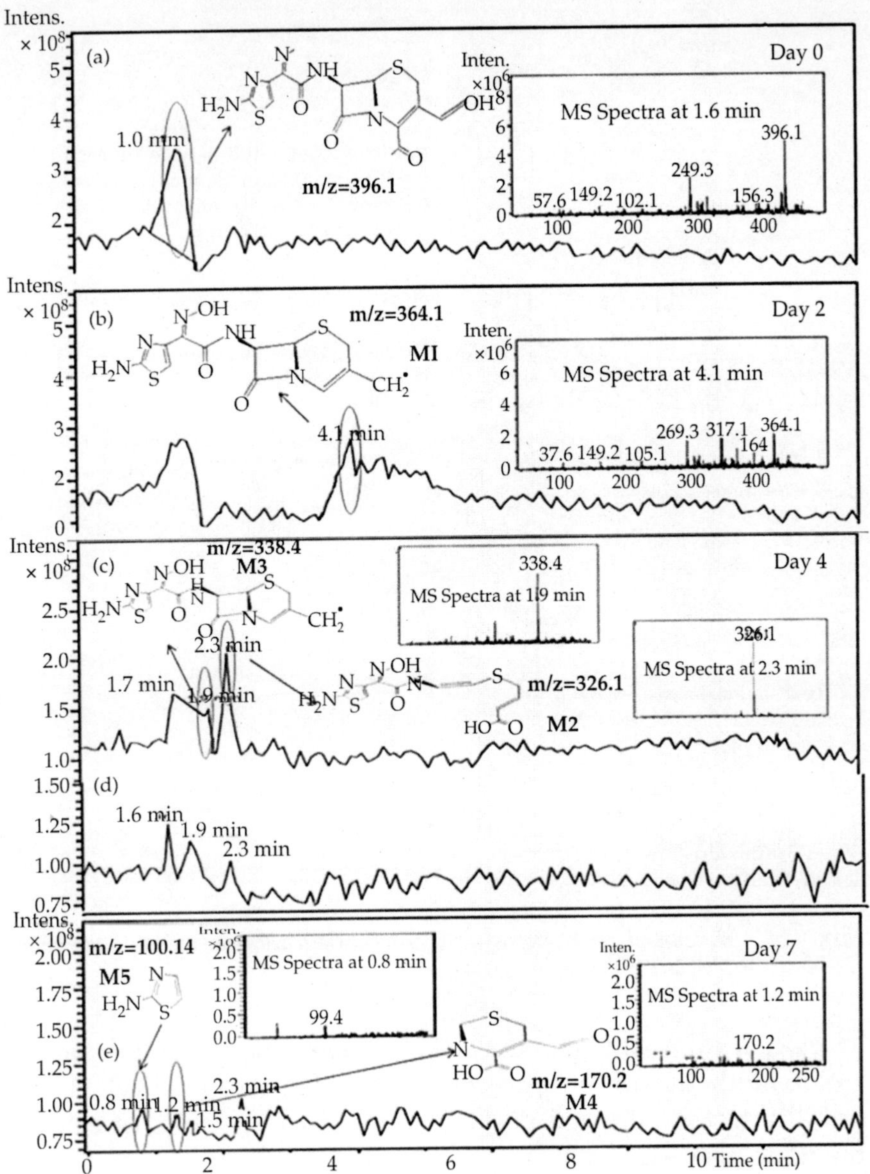

Fig. 9.2: Metabolite characterization by LC–MS. (a) Chromatogram at Day 0 (inset–MS spectra and parent compound structure) (b) Chromatogram at Day 2 (inset–MS spectra, metabolite M1 structure), (c) Chromatogram at Day 4 (inset–MS spectra, metabolite M2 and M3 structure), (d) Chromatogram at Day 6, (e) Chromatogram at Day 7 (inset–MS spectra, metabolite M4 and M5 structure).

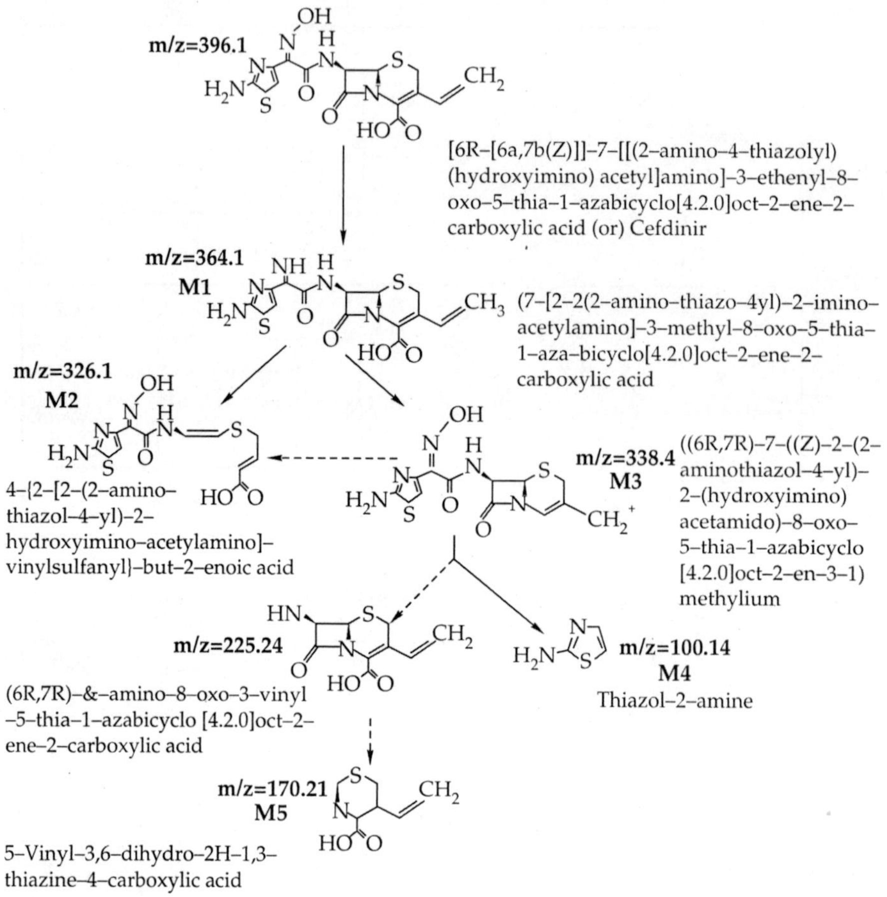

m/z=396.1

[6R–[6a,7b(Z)]]–7–[[(2–amino–4–thiazolyl) (hydroxyimino) acetyl]amino]–3–ethenyl–8–oxo–5–thia–1–azabicyclo[4.2.0]oct–2–ene–2–carboxylic acid (or) Cefdinir

m/z=364.1
M1

(7–[2–2(2–amino–thiazo–4yl)–2–imino–acetylamino]–3–methyl–8–oxo–5–thia–1–aza–bicyclo[4.2.0]oct–2–ene–2–carboxylic acid

m/z=326.1
M2

4–{2–[2–(2–amino–thiazol–4–yl)–2–hydroxyimino–acetylamino]–vinylsulfanyl}–but–2–enoic acid

m/z=338.4
M3

((6R,7R)–7–((Z)–2–(2–aminothiazol–4–yl)–2–(hydroxyimino) acetamido)–8–oxo–5–thia–1–azabicyclo [4.2.0]oct–2–en–3–1) methylium

m/z=225.24

(6R,7R)–&–amino–8–oxo–3–vinyl –5–thia–1–azabicyclo [4.2.0]oct–2–ene–2–carboxylic acid

m/z=100.14
M4

Thiazol–2–amine

m/z=170.21
M5

5–Vinyl–3,6–dihydro–2H–1,3–thiazine–4–carboxylic acid

Fig. 9.3: Proposed pathway of cefdinir biodegradation by *Ustilago sparsa* SMN03 (Selvi *et al.*, 2014).

pharmaceutical wastewater may be considered as a potential microbe for bioremediation of cefdinir–contaminated environments based on its ability to take up and degrade high concentrations of cefdinir from aqueous medium.

BIOSORPTIVE REMOVAL OF ANTIBIOTICS USING YEAST

Among the available process for the treatment of antibiotic containing wastewater, the adsorption process is considered as the most effective

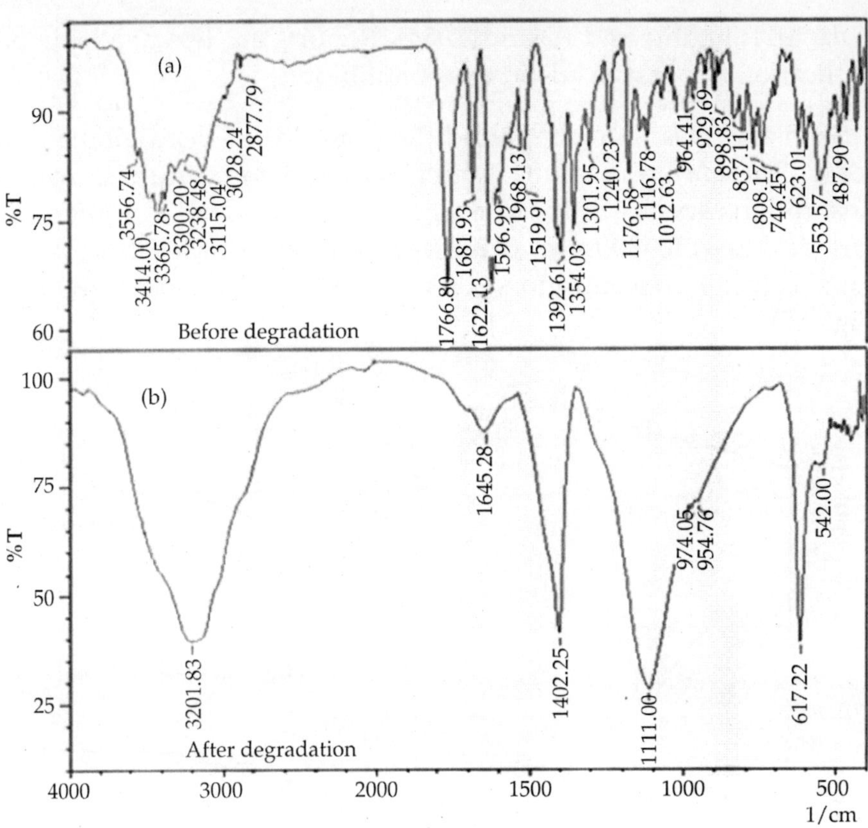

Fig. 9.4: FTIR spectra of cefdinir degradation by *Ustilago sparsa* SMN03. (a) before degradation and (b) after degradation.

and efficient method (Watkinson *et al.*, 2009). Adsorption process has numerous advantages including applicability at very low concentration, suitability for batch and continuous processes, ease of operation, possibility of regeneration and reusability and low capital cost (Delgado *et al.*, 2012). There are reports on adsorption of cephalosporin antibiotics using various adsorbents like activated carbon, nanoparticles, ion–exchange resins etc., which are having drawbacks such as limited success, high cost of the adsorbent, production of toxic compounds, large sludge generation, etc. (Liu *et al.*, 2011; Fakhri and Adami, 2014). Reports are less available on application of yeast as adsorbent of antibiotics.

Selvi and Das (2015) reported the potential of dead biomass of yeast *Candida* sp. SMN04 which showed maximum cefdinir biosorption capacity of 152.67 mg/g under optimized condition *viz.* particle size–150–300 μm, pH–8.0, biosorbent dosage–5% (w/v), initial cefdinir concentration–300 mg/L and contact time–180 min (Fig. 9.5).

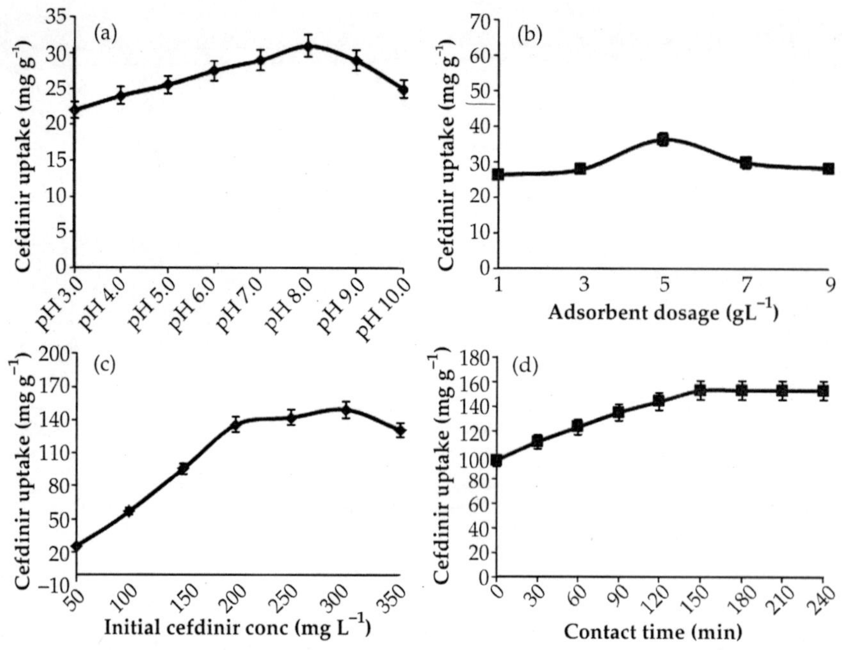

Fig. 9.5: Effect of process parameters on biosorption of cefdinir onto dried *Candida* sp. SMN04. (a) Effect of pH, (b) Effect of biosorbent dosage, (c) Effect of initial concentration and (d) Effect of contact time. Error bars on the curve represent standard deviation of triplicate samples ($p \leq 0.0001$).

The adsorbent (dried biomass of *Candida. Sp.* SMN04) was subjected to various pretreatments. The pretreatment with succinic acid showed maximum cefdinir uptake compared to other pre–treatments (Fig. 9.6). Among the various isotherms tested, Langmuir model fitted best showing maximum adsorption capacity 238.09 mg/g representing monolayer and homogeneous mode of adsorption. The kinetic data were found to follow pseudo–first order model suggesting physisorption as an underlying mechanism (Fig. 9.7).

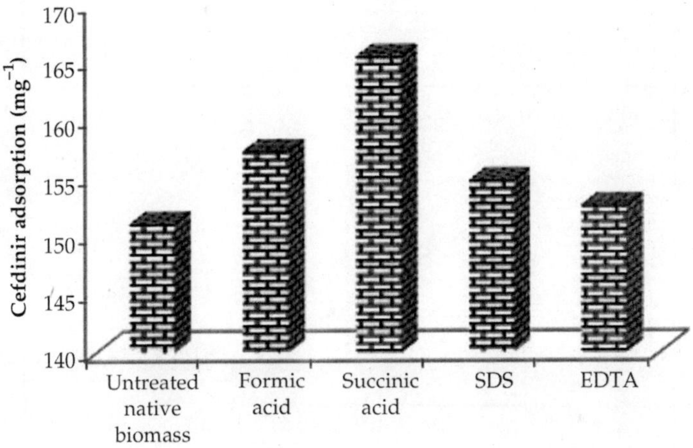

Fig. 9.6: Enhanced cefdinir adsorption by dried *Candida sp.* SMN04 biomass after various pre–treatments (Selvi and Das, 2015).

SEM analysis revealed the surface topology of native (Fig. 9.8a) and succinic acid pre–treated (Fig. 9.8b) dried yeast biomass after cefdinir biosorption. As shown in Fig. 9.8 (a), native biosorbent showed a smooth appearance confirming homogeneous mode of cefdinir adsorption. Pre–treated biosorbent showed porous and irregular surface structure confirming more amount of cefdinir adsorption (Fig. 9.8b).

Based on the reported works as shown in Table 9.3, it was found that succinic acid treated dead yeast biomass could serve as eco–friendly and potential adsorbent for the remediation of cephalosporin antibiotics from aqueous environments.

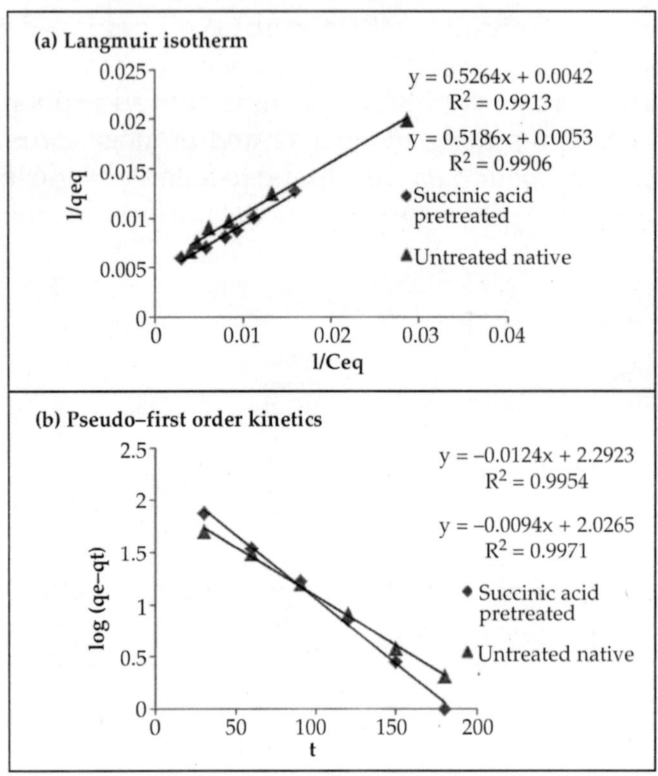

Fig. 9.7: (a) Langmuir isotherm model and (b) Pseudo–first order kinetic model for cefdinir removal using untreated native and succinic acid pre–treated dried *Candida* sp. SMN04 (Selvi and Das, 2015).

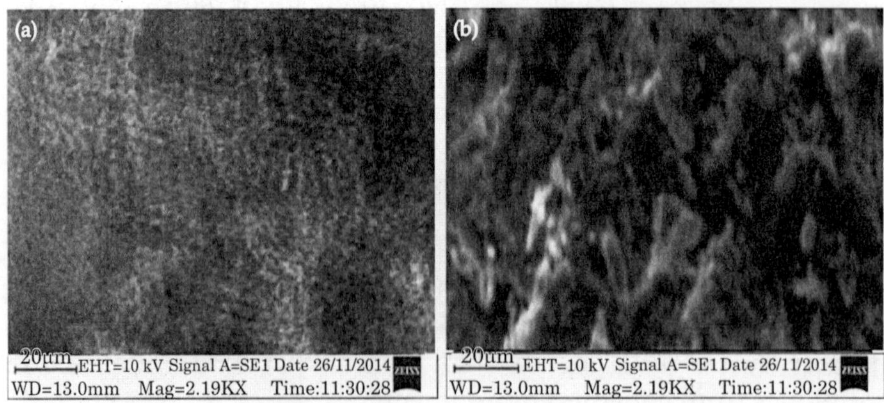

Fig. 9.8: SEM analysis (a) untreated native and (b) succinic acid pre–treated dried *Candida* sp. SMN04 after cefdinir biosorption (Selvi and Das, 2015).

Table 9.3: Reported works on adsorption of pharmaceutical compounds including cephalosporin antibiotics.

Pharmaceutical	*Adsorbents*	*Initial conc of Antibiotics (mg/L)*	*Removal (%)*	*Referen-ce(s)*
Cefradine, Cefalexin, Ceftazidime, Cefixime	Alga–activated sludge	100	89.9 94.9 89.7 100.0	Guo and Chen, 2015
Sulfamethoxazole, Carbazepine,	Mixed microbial culture	100	73.2 4.2	Vasiliadou *et al.*, 2011
Cephalexin	Metal ion modified Activated carbon	4–32	100.0	Liu *et al.*, 2011
Cefixime	MgO nanoparticle	300	31.3	Fakhri and Adami, 2013
Cefdinir	Dried succinic acid Treated yeast biomass	300	81.0	Selvi and Das, 2015

10

Conclusions

The population explosion in the world has resulted an increase in areas of polluted soil and water. As human population is increasing day by day, it creates a growing pressure on our natural resources *i.e.*, air, water and land resources. In order to outfit the demands of the people, the rapid expansion of industries *viz.* food, health care, vehicles, etc. is necessary. But it is very difficult to maintain the quality of life with all these new developments. In nature, there are various microorganisms which are constantly working to break down organic compounds but when pollution occurs, there is a need to clean up. Since the quality of life is inextricably linked to the overall quality of the environment, global attention has been focused on the ways to sustain and preserve the environment. This endeavour is possible by involving technology.

Reports are available on major incidents in the past few decades such as the Exxon Valdez oil spill, Minimata disease in Japan, the Union–Carbide (Dow) Bhopal disaster, large–scale contamination of Rhine river, release of radioactive material in Chernobyl accident and progressive deterioration of aquatic habitats and conifer forests in the Northeastern US, Canada and parts of Europe which have shown the necessity to protect the environment from pollution. For the clean up of contaminated sites, various conventional techniques have been used. Landfilling is a technique which removes the contaminated soil to a landfill or covers the contaminated sites. But this may create significant risks during the process of excavation, handling, and transport of hazardous material. Moreover, it is expensive and very difficult method to find new landfill sites for the final disposal of material. Some other technologies such as high–temperature incineration and chemical use for decomposition have also been applied. These techniques can be effective in reducing level of wide range of contaminants (namely chlorinated solvents, petroleum, polynuclear aromatic hydrocarbons, ketones, PAHs, TNT, RDX, HMX, BTEX, inorganic nitrogen (NO_3, NH_4), explosives, pesticides, herbicides and heavy metals). But these techniques are also having some demerits such as technical complexity and increased contaminants exposure to site workers and nearby residents and lack public acceptance. Various other physico–chemical treatments such as coagulation with alum or lime followed by adsorption on powdered activated carbon (PAC) is reported to yield high removal efficiency for phenolics and COD. These processes generate large volume of hazardous sludge and do not lead to ultimate destruction of the pollutants.

To overcome these problems, a much better perspective is to completely destroy the pollutants, or to transform them into some biodegradable substances. This approach can be achieved by using a technology known as bioremediation. This acts as an option to clean environment and its resources by destroying various contaminants using natural biological activity. Bioremediation is considered as one of the safer, cleaner, cost effective and an environmental friendly technology for decontaminating sites which are contaminated with wide range of pollutants. The process of bioremediation uses various agents such as bacteria, yeast, fungi, algae and higher plants as major tools in treating pollutants present in the environment. In this book, a major thrust has been given on the role of yeasts as remediation agents.

Yeasts are readily available source of biomass which shows the ability to resist under unfavorable environment. Many metals and metalloids can be accumulated by yeasts and some of them are essential for structural and catalytic functions, whereas others are of no metabolic importance. Microbial biomass derived from yeasts, particularly *Candida utilis* has the ability to accumulate metal ions and radionuclides from the environment. The yeast cells of *S. cerevisiae* treated with hot alkali were capable of accumulating a wide range of heavy metal cations (Fe^{3+}, Cu^{2+}, Cr^{3+}, Hg^{2+}, Pb^{2+}, Cd^{2+}, Co^{2+}, Ni^{2+} and Fe^{2+}. The yeast *C. utilis* biomass can conveniently be used for cadmium biosorption from aqueous solutions. The yeast like *Rhodotorula mucilaginosa* is efficient in bioadsorption of lead and also known to accumulate free and complexed silver ions by metabolism dependent an independent processes. The sensitivity of yeast *Pichia guilliermondii* to Cr (III) and Cr (VI) as well as on the uptake potential of Cr was also reported. The results indicated the accumulation of Cr (III) and Cr (IV) by *Pichia* sp. and also showed increase in Cr tolerance by the addition of riboflavin. Immobilized biomass of *Candida utilis* and *Candida tropicalis* showed their efficiencies for the enhanced biosorption of zinc from aqueous solution. Removal of Zn (II) by two yeast isolates *Candida* sp. VITGBN1 and *Cryptococcus* sp. VITGBN2 were also noted. Recent report shows the uranium uptake by both live and heat–killed *Saccharomyces cerevisiae* at environmentally relevant uranium concentration and with different ionic strengths.

Yeasts are also known for decolourization of dye bearing industry effluents mainly by three mechanisms such as biosorption, bioaccumulation and biodegradation. The removal mechanisms of dye color from industrial effluents by yeast cells are either through absorption or adsorption at the cell surface. The yeast species such as *Candida, Clavispora, Debaryomyces, Leucosporidium, Pichia, Rhodosporidium, Rhodotorula, Sporidiobolus, Sporobolomyces, Stephanoascus, Trichosporon* and *Yarrowia* are used in bioremediation process which show biodegradation properties. Baker's yeast, *Saccharomyces cerevisiae* was used as biosorbent for Astrazone Blue basic dye from an aqueous solution. Adsorption of C.I. Reactive Orange 16 (RO 16) from aqueous solution onto residual brewery yeast was investigated. Bioaccumulation of reactive textile dyes Remazol Blue, Remazol Black B and Remazol Red RB by *Candida tropicalis* growing in molasses medium was reported. Combined effects of sugarcane bagasse extract and synthetic dyes on the growth and bioaccumulation properties of yeasts *P. fermentans* MTCC 189 and *C. tropicalis* was reported. Removal of two basic dyes namely Basic Green 4 and Basic Yellow 2 by the yeast *Saccharomyces cerevisiae* was studied in a culture medium prepared from sucrose. Two novel yeast strains *viz. Pseudozyma rugulosa* Y–48 and *Candida krusei* G–1 were reported to be capable of decolorizing multiple types of dyes especially azo dyes which were largely attributable to biodegradation. Biodegradation of an anthraquinone dye, C.I. Disperse Red 15 (DR 15) by yeast strain *Pichia anomala* was reported. Decolorization of an azo dye, Reactive Black 5 by yeast isolate, *Debaryomyces polymorphus* was noted. *Candida oleophila,* a non conventional ascomycetous wild yeast isolate was identified as an excellent degrader of diazo dye Reactive Black 5. The effective biodegradation of malachite green, a triphenylmethane dye by *S. cerevisiae* MTCC 463 was reported.

Biodegradation of the Direct Violet 51 azo dye by *Candida albicans* isolated from industrial effluents was noted. *Trichosporon beigelii* (NCIM–3326), a yeast strain isolated from the tree *Phoebe porphyria* was able to decolorize and degrade reactive azo dye Navy blue HER effectively into non toxic metabolites. Maximum biodegradation of Basic Violet 3 was reported using yeast *C. krusei* when sucrose was used as cosubstrate. Entrapped dead biomass of *Candida tropicalis* in sodium alginate has been used as an efficient dye remediation agent for the treatment of textile wastewater.

Besides being used for dye removal, yeasts are also finding their applications towards remediation of diesel oil, a product of petroleum industry which is considered as major pollutant. Two yeast strains *viz. Saccharomyces cerevisiae* and *Candida albicans* were able to grow effectively on diesel utilizing it as sole sources of carbon and energy. Five yeast species *viz. Candida tropicalis, Cryptococcus laurentii, Trichosporon asahii, Candida rugosa* and *Rhodotorula mucilaginosa* isolated from petroleum–contaminated soil in India were found to be the potent degraders of diesel oil. Recently bio–removal of diesel oil has been reported using a mixed microbial consortium consisted of *Rhodotorula aurantiaca* and *Candida ernobii* alongwith other bacterial species isolated from a polluted environment. There is report on diesel oil degradation by *Candida tropicalis* immobilized on various conventional matrices (sodium alginate, carboxyl methyl cellulose, chitosan) and biowaste materials (wheat bran, sawdust, peanut hull powder). The yeast biofilm formed on gravels by *C. tropicalis* could show 97% degradation of diesel oil over a period of 10 days.

Lindane, a toxic organochlorine pesticide is being used extensively for the control of agricultural and medical pests. The continuous use and indiscriminate industrial production of lindane has created the widespread occurrence of lindane–contaminated soils in the country. Reports are scanty regarding the application of yeast for lindane remediation. Four yeast strains *viz. Pseudozyma* VITJzN01, *C. sorghi* VITJzN02, *Rhodotorula* VITJzN03 and *Candida* VITJzN04 isolated from Indian agricultural soil samples have been reported as potent degraders of lindane. Reported work on degradation of lindane occurred through bio–nano hybrid system involving photocatalytic degradation of lindane using n–ZnO and biodegradation of lindane by *Candida* VITJzN04 in aqueous medium is a novel integrated approach which will serve as an effective remediation tool for the treatment of wastewater and ground waters containing lindane.

Caffeine has been considered as an emerging pollutant in recent years and detected in soil, ground water, surface water and also in wastewater. Efforts have been given towards the remediation of caffeine using yeast to make the various water sources and soil free from this xenobiotic. Five yeast species *viz. Trichosporon asahii, Candida tropicalis, Candida krusei, Candida inconspicua* and *Kloeckera japonica* isolated from caffeine contaminated samples in India were screened for their capability of caffeine degradation. *Trichosporon asahii* showed 60% degradation of caffeine in 72 hours when caffeine was utilized as the sole carbon and nitrogen source. *Saccharomyces cerevisiae* TFS9 isolated from the cultivated tea soils collected from sites of northern Iran showed potentiality of caffeine degradation. This was the first evidence on caffeine bio–degradation using yeast *S. cerevisiae.* Removal of caffeine about 98.3% after 24 days in the biostimulated soil which was bioaugmented with sawdust immobilized yeast mixed culture consisting of three yeast species *viz. Trichosporon asahii, Candida tropicalis* and *Candida inconspicua* was reported.

Antibiotics are unique category of pollutants which are being used extensively in human and veterinary medicine. Pharmaceutical industries involved in the production of antibiotics discharge their wastes openly which contain some quantity of active compounds which are toxic in nature. Globally it is estimated that one million tons of antibiotics are consumed every year and are known to be present at a concentration ranging from few micrograms to kilograms in water and soils posing a threat to the environment. Residues of antibiotics have been detected in surface waters, ground waters, sea waters, drinking water WWTPs effluents, hospital wastewaters and in soil due to application of manure or sludge as fertilizers to the soil. The accumulation and persistence of antibiotics in the environment produce harmful effects, either in aquatic or terrestrial ecosystems. The effluents released from cephalosporin production units were reported to release harmful compounds which are resistant to biodegradation, photo–transformation and natural degradation. The presence of high concentration of cephalosporin in the environment leads to very high chemical oxygen demand thus by increasing the toxic strength of the effluent. Although less work has been done, role of yeasts for remediation of cephalosporin antibiotics have been reported here.

Therefore, based on all the reported works, it can be concluded that yeasts can definitely serve as potential agents similarly like bacteria and fungi for the remediation of various pollutants from soil and water environment.

Bibliography

Abdelwahab, O., N.K. Amin and E–S.Z. El–Ashtoukhy (2013). Removal of zinc ions from aqueous solution using a cation exchange resin. *Chem. Eng. Res. Design*, Vol. 91, pp. 165–173.

Abhilash, P.C. and N. Singh (2010). *Withania somnifera*–dunal mediated dissipation of lindane from simulated soil: Implications for rhizoremediation of contaminated soil. *J. Soils Sediments*, Vol. 10, pp. 272–282.

Ademorati, C.M.A. (1996). Environmental Chemistry and Toxicology. Pollution by Heavy metals, Foludex Press, Ibadan. pp. 171–172.

Agency for Toxic Substances and Disease Registry (ATSDR) (1995). Toxicological for fuel oils. Atlanta, GA: U.S. Department of Health and Human Services, Public Health Service.

Agnihotri, N.P., G. Kulshreshtha, V.T. Gajbhiye, S.P. Mohapatra and S.B. Singh (1996). Organochlorine insecticide residues in agricultural soils of the Indo–Gangetic plains. *Environ. Monit. Assessment*, Vol. 40, pp. 279–288.

Aguirre, J.L., E. Pongráczb, P. Perämäkic and R. L. Keiskia (2010). Micellar–enhanced ultrafiltration for the removal of cadmium and zinc: Use of response surface methodology to improve understanding of process performance and optimization. *J. Hazard. Materials*, Vol. 180, pp. 524–534.

Ahmad, M.F., S. Haydar and T.A. Quraishi (2013). Enhancement of biosorption of zinc ions from aqueous solution by immobilized *Candida utilis* and *Candida tropicalis* cells. *Int. Biodeterior. Biodegradation*, Vol. 83, pp. 119–128.

Ahn, C.K., D. Park, S.H. Woob and J.M. Park (2009). Removal of cationic heavy metal from aqueous solution by activated carbon impregnated with anionic surfactants. *J. Hazard. Materials*, Vol. 164, pp. 1130–1136.

Aksu, Z. (2003). Reactive dye bioaccumulation by *Saccharomyces cerevisiae, Process Biochemistry*, Vol. 38, pp. 1437–1444.

Aksu, Z. (2005). Application of biosorption for the removal of organic pollutants: a review. *Process Biochemistry*, Vol. 40, pp. 997–1026.

Aksu, Z. and G. Bulbul (1999). Determination of the effective diffusion coefficient of phenol in Ca–alginate–immobilized *P. putida* beads–engineering principles, *Enzyme Microb. Technology*, Vol. 25, pp. 344–348.

Aksu, Z. and G. Donmez (2003). A comparative study on the biosorption characteristics of some yeasts for remazol blue reactive dye. *Chemosphere*, Vol. 50, pp. 1075–1083.

Aljendeel, H.A. (2011). Removal of heavy metals using reverse osmosis. *J. Engineering*, Vol. 17, pp. 647–658.

Alkorta, I., J. Hernández–Allica, J.M. Becerril, I. Amezaga, I. Albizu and C. Garbisu (2004). Recent findings on the phytoremediation of soils contaminated with environmentally toxic heavy metals and metalloids such as zinc, cadmium, lead, and arsenic. *Rev. Environ. Sci. Biotechnology*, Vol. 3, pp. 71–90.

Al–Saraj, M., M.S. Abdel–Latif, I. El–Nahal and R. Baraka (1999). Bioaccumulation of some hazardous metals by sol–gel entrapped microorganisms. *J. Non–Cryst. Solids*, Vol. 248, No. 2, pp. 137–140.

Altschul, S.F., W. Gish, W. Miller, E.W. Myers and D.J. Lipman (1990). Basic local alignment search tool. *J. Mol. Biology*, Vol. 5, pp. 403–410.

Anaya, A.L., G.R. Waller, P.O. Owuor, J. Friedman, C.H. Chou, T. Suzuki, J.F. Arroyo–Estrada and R. Cruz–Ortega (2002). The role of caffeine in the production decline due to autotoxicity in coffee and tea plantations, *In*: Allelopathy from molecules to ecosystems, M. Reigosa, N. Pedrol (*ed.*), Science Publishers, Enfield, pp. 71–92.

Anderson, R.A. (1998). Effects of chromium on body composition and weight loss. *Nutr. Rev.*, Vol. 56, No. 9, pp. 266–270.

Andreozzi, R., M. Canterino, R. Marotta and N. Paxeus (2005). Antibiotic removal from wastewaters: The ozonation of amoxicillin. *J. Hazard. Materials*, Vol. 122, pp. 243– 250.

Arikan, O.A. (2008). Degradation and metabolization of chlortetracycline during the anaerobic digestion of manure from medicated calves. *J. Hazard. Materials*, Vol. 158, pp. 485–490.

Armenta, S., S. Garrigues and M. de la Guardia (2005). Solid–phase FT–Raman determination of caffeine in energy drinks. *Anal. Chim. Acta*, Vol. 547, pp. 197–203.

Arslan I.A., A.B. Isil and W.B. Detlef (2002). Advanced oxidation of a reactive dyebath effluent: comparison of O_3, H_2O_2/UV–C and TiO_2/UV–A processes, *Wat. Research*, Vol. 36, pp. 1143–1154.

Arslan–Alaton, I. and A.E. Caglayan (2006). Toxicity and biodegradability assessment of raw and ozonated procaine penicillin G formulation effluent. *Ecotoxicol. Environ. Safety*, Vol. 63, pp. 131–140.

Arthur, H. and K. Watson (1976). Thermal adaptation in yeast: Growth temperatures, membrane lipid, and cytochrome composition of psychrophilic, mesophilic, and thermophilic yeasts. *J. Bacteriology*, Vol. 128, pp. 56–68.

Ashengroph, M. and M Borchaluei (2013). *Saccharomyces cerevisiae* TFS9, a novel isolated yeast capable of high caffeine–tolerant and its application in biodecaffeination approach. *Prog. Biol. Sciences* ,Vol. 3, No. 2, pp. 145–156.

Ashihara, H., and A. Crozier (2001). Caffeine: A well known but little mentioned compound in plant science. *Trends Plant Science,* Vol. 6, pp. 407–413.

Asmatullah, S.N. Qureshi and A.R. Shakoori (1998). Hexavalent chromium induced congenital abnormalities in chick embryos. *J. Appl. Toxicology*, Vol. 18, No. 3, pp. 167–171.

Atlas R. and J. Bragg (2009). Bioremediation of marine oil spills: When and when not the Exxon Valdez experience. *Microb. Biotechnology,* Vol. 2, pp. 213–221.

ATSDR (2005). US Department of Health and Human Services, toxicological profile for alpha–, beta–, gamma–, and delta– hexachlorocyclohexane, *Agency for Toxic Substances and Disease Registry,* http://www.atsdr.cdc.gov/toxprofiles/tp43.pdf.

Aurnaud, M.J. (1987). The pharmacology of caffeine. *Prog. Drug Research,* Vol. 31, pp. 273–313.

Awasthi, N., R. Ahuja and A. Kumar (2000). Factors inuencing the degradation of soil applied endosulfan isomers. *Soil Biol. Biochemistry,* Vol. 32, pp. 1697–1705.

Babic, S., A.J.M. Horvat, D.M. Pavlovic and M. Kastelan–Macan (2007). *Trends Anal Chemistry,* Vol. 26, pp. 1043–1061.

Babu, V.R.S, S. Patra, M.S. Thakur, N.G. Karanth, and M.C. Varadaraj (2005). Degradation of caffeine by *Pseudomonas alcaligenes* CFR 1708, *Enzyme Microb. Technology.* Vol. 37, pp. 617–624.

Bakalar, T., M. Bugel and L. Gajdosova (2009). Heavy metal removal using reverse osmosis. *Acta Mont. Slovaca,* Vol. 14, pp. 250–253.

Bakkaloglu, I., T.J. Butter, L.M. Evison, F.S. Holland and I.C. Hancock (1998). Screening of various types of biomass for removal and recovery of heavy metals (Zn, Cu, Ni) by biosorption, sedimentation and desorption. *Water Sci. Technology,* Vol. 38, pp. 269–277.

Bakore, N., P.J. John and P. Bhatnagar (2004). Organochlorine pesticide residues in wheat and drinking water samples from Jaipur, Rajasthan, India. *Environ. Monit. Assessment,* Vol. 98, pp. 381–389.

Banat, I.M., P. Nigam, D. Singh and R. Marchant (1996). Microbial decolorization of textile–dye containing effluents: A review, *Bioresour. Technology,* Vol. 58, pp. 217–227.

Bankar, A.V., A.R. Kumar and S.S. Zinjarde (2009a). Environmental and industrial applications of *Yarrowia lipolytica, Appl. Microbiol. Biotechnology,* Vol. 84, pp. 847–865.

Bankar, A.V., A.R. Kumar and S.S. Zinjarde (2009b). Removal of chromium (VI) ions from aqueous solution by adsorption onto two marine isolates of *Yarrowia lipolytica. J. Hazard. Materials,* Vol. 170, pp. 487–494.

Bantle, J.A, D.T. Burton, D.A. Dawson, J.N. Dumont, R.A. Finch, D.J. Fort, *et al.* (1994). FETAX interlaboratory validation study: Phase II testing, *Environ. Toxicol. Chemistry,* Vol. 13, pp. 1629–1637.

Barathi, S. and N. Vasudevan (2003). Bioremediation of crude oil contaminated soil by bioaugmentation of *Pseudomonas fluorescens* NS1, *J. Environ. Sci. Health. A,* Vol. 38, pp. 1857–1866.

Basak, G., D. Charumathi and N. Das (2011). Combined effects of sugarcane bagasse extract and zinc (II) ions on the growth and bioaccumulation properties of yeast isolates. *Int. J. Eng. Sci. Technology,* Vol. 3, pp. 6321–6334.

Basak, G. and N. Das (2013). Zinc(II) removal by chemically treated dead biomass of yeast species. *Nat. Environ. Pollut. Technology,* Vol. 12, pp. 81–86.

Basak, G. and N. Das (2014). Characterization of sophorolipid biosurfactant produced *by Cryptococcus* sp. VITGBN2 and its application on Zn(II) removal from electroplating wastewater. *J Env. Bio*logy, Vol. 35 , No. 6 , pp. 1087–1094.

Basak, G., V. Lakshmi, P. Chandran and N. Das (2014 a). Removal of Zn(II) from electroplating effluent using yeast biofilm formed on gravels: Batch and column studies. *J. Environ. Health. Sci. Engineering*, Vol. 12, pp. 8–11., doi: 10.1186/ 2052–336X–12–8.

Basak, G., D. Das and N. Das (2014b). Dual role of acidic diacetate sophorolipid as biostabilizer for ZnO nanoparticle synthesis and biofunctionalizing agent against *Salmonella enterica* and *Candida albicans*, *J. Microbiol. Biotechnology*, Vol. 24, pp. 87–96.

Baskaran, V. and M. Nemati (2006). Anaerobic reduction of sulfate in immobilized cell bioreactors using a microbial culture originated from an oil reservoir, *Biochem. Engineering*, Vol. 31, pp. 148–159.

Batish, D.R., H.P. Singh, M. Kaur, R.K. Kohli and S.S. Yadav (2008). Caffeine affects adventitious rooting and causes biochemical changes in the hypocotyl cuttings of mung bean (*Phaseolus aureus* Roxb.), *Acta Physiol. Plant*, Vol. 30, pp. 401–405.

Batt, A.L., D.D. Snow and D.S. Aga (2006). Occurrence of sulphonamide antimicrobials in private water wells in Washington County, Idaho, USA, *Chemosphere*, Vol. 64, pp. 1963–1971.

Baysal, Z., E. Cinar, Y. Bulut, H. Alkan and M. Dogru (2009). Equilibrium and thermodynamic studies on biosorption of Pb (II) onto *Candida albicans* biomass. *J. Hazard. Mate*rials, Vol. 161, pp. 62–67.

Benimeli, C.S., G.R. Castro, A.P. Chaile and M.J. Amoroso (2007) Lindane uptake and degradation by aquatic *Streptomyces* sp. strain M7, *Int. Biodeter. Biodegradation*, Vol. 59, pp. 148–155.

Bennett, J.W. and B.D. Faison (1997). Use of fungi in biodegradation. *In*: Hurst C.J., G.R. Knudsen, M.J. Mclnerney, L.D. Stetzenbach, M.V. Walter (*eds.*), Manual of Environmental Microbiology, ASM Press, pp. 758–765, Washington DC.

Bernabeu, A., R.F. Vercher, L. Santos–Juanes, P.J. Simón, C. Lardín, M.A. Martínez, J.A. Vicente, R. González, C. Llosá, A. Arques and A.M. Amat (2012). Solar photocatalysis as a tertiary treatment to remove emerging pollutants from wastewater treatment plant effluents, *Catalysis Today* , Vol. 161, pp. 235–240.

Beyer, A. and M. Matthies (2001). Long–range transport potential of semi–volatile organic chemicals in coupled air–water systems, *Environ. Sci. Pollut. Reseasrch*, Vol. 8, No. 3, pp. 173–179.

Bhattacharya, B., S.K. Sarkar and N. Mukherjee (2003). Organochlorine pesticide residues in sediments of a tropical mangrove estuary, India: Implications for monitoring. *Environ. International*, Vol. 29, pp. 587–592.

Blakley, B.R. (1982). Lindane toxicity in pigeons, *Can. Vet. Journal*, Vol. 23, pp. 267–268.

Blanquez, P., N. Casas, X. Font, X. Gabarrell, M. Sarra, G. Caminal and T. Vicent (2004). Mechanism of textile metal dye biotransformation by *Trametes versicolor*, *Water Research*, Vol. 2, pp. 166–2172.

Bolong, N., A.F. Ismail, M.R. Salim and T. Matsuura (2009). A review of the effects of emerging contaminants in wastewater and options for their removal, *Desalination*, Vol. 239, pp. 229–246.

Boopathy, R (2000). Factors affecting bioremediation technologies, *Bioresour. Technology*, Vol. 74, pp. 63.

Botstein, D. and G.R. Fink (2011). Yeast: an experimental organism for 21st Century biology, *Genetics*, Vol. 189, pp. 695–704.

Brady, J.M. and J.M. Tobin (1994). Adsorption of metal ions by *Rhizopus arrihizus* biomass: Characterization studies, *Enzyme Microb. Technology*, Vol. 6, pp. 671–675.

Brat, D., E. Boles and B. Wiedemann (2009). Functional expression of a bacterial xylose isomerase in *Saccharomyces cerevisiae*, *Appl. Environ. Microbiology*, Vol. 75, No. 8, pp. 2304–2311.

Britto, J.M. and M.C. Rangel (2008). Processos avançados de oxidação de compostos fenólicos em efluentes industriais. *Quím. Nova* Vol. 31, 114–122.

Brunner, G.H. (1994). Extraction and destruction of waste with supercritical water. NATO ASI Ser., Ser. E. 273 (Supercritical fluids), pp. 697–705.

Buerge, I.J.,.Poiger, M.D. Muller and H.R. Buser (2003). Caffeine, an anthropogenic marker for wastewater contamination of surface waters, *Environ. Sci. Technology*, Vol. 37, pp. 691–700.

Bustard M., G. Mc. Mullan and A.P. Mc Hale (1998). Biosorption of textile dyes by biomass derived from *Kluyveromyces marxianus* IMB3, *Bioprocess Engineering*, Vol. 19, pp. 427–430.

Butinar L, Santos S Spencer–Martins I, Oren A and Gunde–Cimerman N.(2005) Yeast diversity in hypersaline habitats, *FEMS Microbiol Letters*. 2005, Vol. 15, No. 244(2), pp. 229–34.

Camacho–Pérez, B., E. Ríos–Leal, N. Rinderknecht–Seijas and H.M. Poggi–Varaldo (2012). Enzymes involved in the biodegradation of hexachlorocyclohexane: A mini review, *J Environ Management*, Vol. 95, pp. S306–S318.

CARB (2008). California Air Resources. Staff Report: Methodology for Estimating Premature Deaths Associated with Long–term Exposure to Fine Airborne Particulate Matter in California, http://www.arb.ca.gov/research/health/pm–mort/pm–mort_final.pdf. 2008.

Cardoso,O., J.M. Porcher and W. Sanchez (2014) Factory discharged pharmaceuticals could be a relevant source of aquatic environment contamination; Review of evidence and need for knowledge, *Chemosphere* Vol. 115, pp. 20–30.

Carracciolo, A.B., E.Topp and P. Grenni (2015) Pharmaceuticals in environment: Biodegradation and effects on natural microbial communities, A Review, *J. Phar. Biomed. Analysis*, Vol. 106, pp. 25–36.

Carraway, E.R., A.F. Hoffmann and M.R. Hoffmann (1994). Photocatalytic oxidation of organic acids on quantum-sized semiconductor colloids, *Environ. Sci. Technology*, Vol. 28, pp. 786–785.

Casqueira, R.G, M.L. Toremb and H.M. Kohlerb (2006). The removal of zinc from liquid streams by electroflotation, *Miner. Engineering*, Vol. 19, pp. 1388–1392.

Cazenave, J. D.A. Wunderlin, M.A. Bistoni, M.V. Ame, C. Wiegand, E. Krause and S. Pflugmacher (2005). Uptake, tissue distribution and accumulation of microcystin–RR in *Corydoras paleatus, Jenynsia multidentata* and *Odontesthes bonariensis*: A field and laboratory study, *Aquat. Toxicol*ogy, Vol. 75, No. 2, pp. 178–190.

Cha, C.J, D.R. Doerge and C.E.Cerniglia (2001). Biotransformation of malachite green by the fungus *Cunninghamella elegans, Appl. Environ. Microbiology*, Vol. 67, pp. 4358–4360.

Chalivendra, S. (2011). Catalytic destruction of lindane using a nano iron oxide catalyst. *PhD diss., University of Dayton.*

Chandran, P. and N. Das (2010). Biosurfactant production and diesel oil degradation by yeast species *Trichosporon asahii* isolated from petroleum hydrocarbon contaminated soil, *Int. J. Eng. Sci. Technology*, Vol. 2, pp. 6942–6953.

Chandran, P. and N. Das (2011). Degradation of diesel oil by immobilized *Candida tropicalis* and biolm formed on gravels, *Biodegradation,* Vol. 22, pp. 1181–1189.

Chandran, P. and N. Das (2012). Role of plasmid in diesel oil degradation by yeast species isolated from petroleum hydrocarbon–contaminated soil, Environ. Technology, Vol. 33, No. (4–6), pp. 645–652.

Charumathi, D. and N. Das (2010). Removal of synthetic dye Basic Violet 3 by immobilised *Candida tropicalis* grown on sugarcane bagasse extract medium, *Int. J. Eng. Science Technology*, Vol. 2, No. 9, pp. 4325–4335.

Charumathi, D. and N. Das (2011). Biodegradation of Basic Violet 3 by *C.krusei* isolated from textile wastewater, *Biodegradation*, Vol. 22 , pp. 1169–1180 .

Charumathi,D. And N. Das (2012) Packed bed column studies for the removal of synthetic dyes from textile wastewater using immobilised dead *C. tropicalis*, Desalination, Vol. 285, pp. 22–30.

Chatterjee, S.K., I. Bhattacharjee and G. Chandra (2010). Biosorption of heavy metals from industrial waste water by *Geobacillus thermodenitrificans, J. Hazard. Materials*, Vol. 175, pp. 117–125.

Chen, C.C., H.J. Liao, C.Y. Cheng, C.Y. Yen and Y.C. Chung (2007). Biodegradation of crystal violet by *Pseudomonas putida, Biotechnol. Letters*, Vol. 29, pp. 391–396.

Cheng, H.R. and N. Jiang (2006). Extremely rapid extraction of DNA from bacteria and yeast, *Biotechnol. Letters*, Vol. 28, pp. 58–59.

Cho, B.P., T. Yang, L.R. Blankenship, J.D. Moody, M. Churchwell *et al.* (2003). Synthesis and characterization of N–demethylated metabolites of malachite green and leuco malachite green, *Chem. Res. Toxicology*, Vol. 16, pp. 285–294.

Choi, H., K. Zhang, D.D. Dionysiou, D.B. Oerther and G.A. Sorial (2005). Effect of permeate flux and tangential flow on membrane fouling for wastewater treatment, *Sep. Purif. Technology*, Vol. 45, pp. 68–78.

Clarke, E.A. and R. Anliker (1980). Organic dyes and pigments, *In*: Handbook of Environmental Chemistry, Anthropogenic Compounds, Part A, Vol. 3, Springer, p. 181–215, New York, 1980 pp. 181–215.

Clewell, H.J. (1981). The effect of fuel composition on groundfall from aircraft fuel jettisoning report. Air Force engineering and services center, Tyndall airforce base, FL.

Cokgor, E.U., I. Arslan–Alaton, O. Karahan, S. Dogruel and D. Orhon (2004). Biological treatability of raw and ozonated penicillin formulation effluent, *J. Hazard. Materials*, Vol. 116, pp. 159–166.

Cunningham, V.I (2008) In: Kummerer K (ed) Pharmaceuticals in the environment. Springer, pp. 23–33.

Dai, C., C.Q. Zheng, M. Jiang, X.Y. Ma and L.J. Jiang (2013).Probiotics and irritable bowel syndrome, *World. J. Gastroenterology*, Vol. 19, pp. 5973–5980.

Dantas, R.F., S. Contreras, C. Sans and S. Esplugas (2008). Sulfamethoxazole abatement means of ozonation. *J. Hazard. Materials* , Vol. 150, pp. 790–794.

Das, D., D. Charumathi and N. Das (2010). Combined effects of sugarcane bagasse extract and synthetic dyes on the growth and bioaccumulation properties of *Pichia fermentans* MTCC 189, *J. Hazard. Materials*, Vol. 183, pp. 497–505.

Das, D., D. Charumathi and N. Das (2011). Bioaccumulation of synthetic dye Basic Violet 3 and heavy metals in single and binary systems by *Candida tropicalis* grown in sugarcane bagasse extract medium: Use of response surface methodology (RSM) and inhibition kinetics, *J. Hazard Mater*, Vol. 186, pp. 1541–1552.

Das, D., G. Basak, V. Lakshmi and N. Das (2012). Kinetics and equilibrium studies on removal of zinc(II) by untreated and anionic surfactant treated dead biomass of yeast: Batch and column mode, *Biochem. Eng. Journal*, Vol. 64, pp. 30–47.

Das, N. and P. Chandran (2011). Microbial degradation of petroleum hydrocarbon contaminants: An overview, *Biotechnol. Res. International*, 10.4061/2011/941810.

Das, N., R. Vimala and P. Karthika (2008). Biosorption of heavy metals–An Overview. *Indian J. Biotechnology*., Vol. 7, pp. 159–169.

Dash, S. S. and S.N. Gummadi (2006a). Biodegradation of caffeine by *Pseudomonas* sp. NCIM 5235. *Res. J. Microbiology*, Vol. 1, pp. 115–123.

Dash, S.S. and S.N. Gummadi (2006b). Catabolic pathways and biotechnological applications of microbial caffeine degradation, *Biotechnol. Let*ters, Vol. 28, pp. 1993–2002.

Dash, S.S. and S.N. Gummadi (2007a). Optimization of physical parameters for biodegradation of caffeine by *Pseudomonas* sp.: A Statistical approach, *Am J Food Technology*, Vol. 2, No. 1, pp. 21–29.

Davis,J.A., B. Volesky and R.H.S.F. Vierra (2000). Sargassum seaweed as biosorbent for heavy metals, *Water Res*earch, Vol. 34, No. 17, pp. 4270–4278.

Dawkar, V.V., U.U. Jadhav, D.P.Thamboli and S.P. Govindwar (2010). Efficient industrial dye decolorization by *Bacillus* sp. VUS with its enzyme system, *Ecotoxic. Environ. Safety*, Vol. 73, pp. 1696–1708.

De Angelis, F.E. and G.S. Rodrigues. (1987). Azo dyes removal from industrial effluents using yeast biomass, *Arquivos De Biologia E. Technologia,* Vol. 30, pp. 301–309.

Decampo, R. and S.N. Moreno (1990). The metabolism and mode of action of gentian violet, *Drug. Metab. Reviews,* Vol. 22, pp. 161–178.

Delgado LF., P. Charles, K. Glucina and C. Morlay (2012). The removal of endocrine disrupting compounds, pharmaceutically activated compounds and cyanobacterail during drinking water preparation using activated carbon–a review. *Science of the Total Environment.*Vol. 435–436, pp. 509–525.

Dermentzis, K., A. Christoforidis and E. Valsamidou (2011). Removal of nickel, copper, zinc and chromium from synthetic and industrial wastewater by electrocoagulation, *Int. J. Environ. Sciences,* Vol. 1, pp. 697–710.

Dinleyici, E.C., M. Eren, M. Ozen, Z.A. Yargic and Y. Vandenplas (2012). Effectiveness and safety of *Saccharomyces boulardii* for acute infectious diarrhea, *Expert. Opin. Biol. Therapy,* Vol. 12, pp. 395–410.

Dolliver, H.A., K. Kumar and S.C. Gupta (2007). Sulfamethazine uptake by plants from manure–amended soil, *J. Environ. Quality,* Vol. 36, pp. 1224 –1230.

Dong, Z.B, Y.R.Liang, F.Y. Fan, J.H. Ye, X.Q. Zheng and J.L. Lu (2011). Adsorption behavior of the catechins and caffeine onto polyvinylpolypyrrolidone, *J. Agric. Food Chemistry,* Vol. 59, pp. 4238–4247.

Donmez G. (2002). Bioaccumulation of the reactive textile dyes by *Candida tropicalis* growing in molasses medium, *Enzyme Microb. Technology,* Vol. 20, pp 363–366.

Drake, L.R., S. Lin, G.D. Rayson, and P.J. Jackson (1996). Chemical modification and metal binding studies of *Datura innoxia, Environ. Sci. Technology,* Vol. 30, pp. 110–114.

Du, L., M. Zhao, G. Li, F. Xu, W. Chen and Y. Zhao (2013). Biodegradation of malachite green by *Micrococcus* sp. strain BD15: Biodegradation pathway and enzyme analysis, *Int. Biodeter. Biodegradation,* Vol. 78, pp. 108–116.

Dyal, A., K. Loos, K, M. Noto, S.W. Chang, C. Spagnoli, K.V. Shafi, A. Ulman, M. Cowman and R.A. Gross (2003). Activity of *Candida rugosa* lipase immobilized on gamma–Fe_2O_3 magnetic nanoparticles, *J. Am. Chem. Society,* Vol. 125, No. 7, pp. 1684–1685.

Eckenfelder, W.W., (2007). Wastewater treatment. *In*: Kirk, E.R., Othmer, D.F., Kroschwitz, J.I., Howe–Grant, M. (*Eds.*), KirkeOthmer Encyclopedia Chemical Technology. John Wiley & Sons, New York.

Ehrenfeld, J.R. and J. Boss (1984). Evaluation of remedial action, in Unit operations at hazardous waste disposal sites. Noyes publications, Park ridge, New Jersey, pp. 435.

El Beit, I.O.D., J.V. Wheelock and D.E. Cotton (1981). Factors aecting soil residues of dieldrin, endosulfan, γ–HCH, dimethoate and pyrolan, *Ecotoxicol. Environ. Safety* Vol. 5, pp. 135–160.

Elcey, C.D. and A.A.M., Kunhi (2010). Substantially enhanced degradation of hexachlorocyclohexane isomers by a microbial consortium on acclimation, *J. Agric. Food Chemistry,* Vol. 58, pp. 1046–1054.

Elimelech, M. and R.O. Charles (1990). Kinetics of deposition of colloidal particles in porous media, *Environ. Sci. Technology*, Vol. 24, pp. 1528–1536.

Ellis, D.E., E.J. Lutz, J.M.Odom, Jr. R.J. Buchanan and C.L. Bartlett (2000). Bioaugmentation for accelerated *in situ* anaerobic bioremediation, *Environ. Sci. Technology*, Vol. 34, pp. 2254–2260.

Elmolla, E and M. Chaudhuri (2009b). Optimization of Fenton process for treatment of amoxicillin, ampicillin and cloxacillin antibiotics in aqueous solution, *J. Hazard. Materials*, Vol. 170, pp. 666–672.

Faisal, M. and S. Hasnain (2004). Microbial conversion of Cr (VI) into Cr (III) in industrial effluent, *Afr. J. Biotechnol*ogy, Vol. 3, No. 11, pp. 610–617.

Fakhri A, and S. Adami S. (2014). Adsorption and thermodynamic studies of cephalosporin antibiotics from aqueous solution onto MgO. *J. Taiwan Inst. Chem. Engineers*, Vol. 45, pp. 1001–1006.

Fan, L., A. Pandey and C.R. Soccol (1999). Proceedings of the Paper Presented at the 3rd International Seminar on Biotechnology in the Coffee Agro–industry, Londrina, Brazil, p. 51.

Fang–Y. F., Y. Xu, Yue–R. Liang, X–Q. Zheng, D Borthakur and J–L. Lu (2011). Isolation and characterization of high caffeine–tolerant bacterium strains from the soil of tea garden, *Afr J Microbiol Research*, Vol. 5, pp. 2278–2286.

Farah, J.Y., N.S. El–Gendy and L.A. Farahat (2007). Biosorption of Astrazone Blue basic dye from an aqueous solution using dried biomass of Baker's yeast, *J. Hazard. Materials*, Vol. 148, pp. 402–408.

Feitosa–Felizzola, J and S. Chiron (2009). Occurrence and distribution of selected antibiotics in a small Mediterranean stream (Arc River, Southern France), *J. Hydrology*, Vol. 364, pp. 50–57.

Feldman, J.R. and S.N. Katz (1997). Caffeine. *In*: Encyclopedia of chemical processing and design Marcel Dekker Inc, pp. 424–440.

Fernandes, O., M. Sabharwal, T. Smiley, A. Pastuszak, G. Koren and T. Einarson (1998). Moderate to heavy caffeine consumption during pregnancy and relationship to spontaneous abortion and abnormal fetal growth: a meta–analysis, *Reprod. Toxicology*, Vol. 12, pp. 435–444.

FNB. (1974). Recommended dietary allowances. Food and nutrition Board, National Academy of Sciences, Washington D.C.

Fomina, M. and G. M. Gadd (2014). Biosorption: Current perspectives on concept, definition and application, *Bioresour. Technol*ogy, Vol. 160, pp. 3–14.

Fox, R.D., E.S. Alperin., H.H. Huls (1991). Thermal treatment for the removal of PCBs and other organics from soil, *Environ. Program*, Vol. 10, pp. 40–44.

Friedman, J. and G.R. Waller (1983). Caffeine hazards and their prevention in germinating seed of coffee, *J. Chem. Ecology*, Vol. 9, pp. 1099–1106.

Gallego, J.L.R.,J. Loredo., J.F. Llamas., F. Vasquez and J. Sanchez (2001). Bioremediation of diesel contaminated soils: Evaluation of potential *in situ* techniques by study of bacterial degradation, *Biodegradation*, Vol. 12, pp. 325–335.

Gartiser, S., E.Urich , R. Alexy and K. Kummerer (2007). Ultimate biodegradation and elimination of antibiotics in inherent tests, *Chemosphere* Vol. 67, pp. 604 613.

Ghosh, P., A.N. Samantha and S. Ray (2011). Reduction of COD and removal of Zn^{2+} from rayon industry wastewater by combined electrofenton treatment and chemical precipitation. *Desalination*, Vol. 266, pp. 213–217.

Gilden, R.C., K. Huffling and B. Sattler (2010). Pesticides and health risks, *J. Obstet. Gynecol. Neonatal. Nursing*, Vol. 39, pp. 103–110.

Gladfelter, A.S. (2006). Nuclear anarchy: Asynchronous mitosis in multinucleated fungal hyphae, *Curr. Opin. Microbiology*, Vol. 9, pp. 547–552.

Glassmeyer, S.T., E.T. Furlong, D.W. Koplin, J.D. Cahill, S.D. Zaugg, S.L. Werner, M.T. Meyer, and D.D. Kryak (2005). Transport of chemical and microbial compounds from known waste water discharges: Potential for use as indicators of human fecal contamination, *Environ. Sci. Technology*, Vol. 39, pp. 5157–5169.

Godbole, P.T. and A.D. Sawant (2006). Removal of malachite green from aqueous solutions using immobilised *Saccharomyces cerevisiae, J. Sci. Ind. research*, Vol. 65, pp. 440–442.

Goksungur, Y., S. Uren and U. Guvenc (2005). Biosorption of cadmium and lead ions by ethanol treated waste baker's yeast biomass, *Bioresour Technology*, Vol. 96, pp. 103–109.

Gokulakrishnan, S., K. Chandraraj and S.N. Gummadi (2005). Microbial and enzymatic methods for the removal of caffeine, *Enzyme Microb. Technology*, Vol. 37, pp. 225–232.

Gomes, N.C.M., C.A. Rosa, P.F. Pimentel and L.C.S. Mendonca–Hagler (2002). Uptake of free and complexed silver ions by different strains of *Rhodotorula mucilaginosa, Braz. J. Microbiology*, Vol. 33, pp. 62–66.

Gonen, F. and Z. Aksu (2009a). Predictive expressions of growth and Remazol turquoise blue–G reactive dye bioaccumulation properties of *Candida utilis, Enzyme Microb. Technology*, Vol. 45, pp. 15–21.

Gonen, F. and Z. Aksu (2009b). Single and binary dye and heavy metal bioaccumulation properties of *Candida tropicalis*: Use of response surface methodology (RSM) for the estimation of removal yields, *J. Hazard. Materials*, Vol. 172, pp. 1512–1519.

Grigg, C. W. (1972). Effects of coumarin, pyronine Y, 69dimethyl 2methylthiopurine and caffeine on excision repair and recombinationin *Escherichia coli, J. Gen. Microbiology*, Vol. 70, pp. 221–230.

Guillén–Jiméneza, F., E. Cristiani–Urbinab, J.C. Cancino–Díazc, J.L Flores–Morenod and B.E. Barragán–Huertaa (2012). Lindane biodegradation by the *Fusarium verticillioides* AT–100 strain, isolated from *Agave tequilana* leaves: Kinetic study and identification of metabolites, *Int. Biodeter. Biodegradation*, Vol. 74, pp. 36–47.

Guinea, E., J.A. Garrido, R.M. Rodríguez, P.L. Cabot, C. Arias, F. Centellas and E. Brillas (2010). Degradation of the fluoroquinolone enrofloxacin by electrochemical advanced oxidation processes based on hydrogen peroxide electrogeneration, *Electrochim. Acta*, Vol. 55, pp. 2101–2115.

Gummadi, S.N. and D. Santhosh (2006). How induced cells of *Pseudomonas* sp. increase the degradation of caffeine, *Cent. Eur. J. Biology*, Vol. 1, pp. 561–571.

Guo, R and J. Chen (2015). Application of alga activated sludge combined system (AASCS) as a novel treatment to remove cephalosporins. *Chem. Eng. J.*, Vol. 260, pp. 550–556.

Gupta, P.K. (1986). Pesticides in the India environment. *Interprint, New Delhi*, pp. 1–206.

Gupta, V.K., C.K. Jain, I. Ali, S. Chandra and S. Agarwal (2002). Removal of lindane and malathion from wastewater using bagasse fly ash—a sugar industry waste, *Water Research*, Vol. 36, pp. 2483–2490.

Hamscher, G., S. Sczesny, H. Hoper and H. Nau (2002). Determination of persistent tetracycline residues in soil fertilized with liquid manure by high–performance liquid chromatography with electrospray ionization tandem mass spectrometry, *Anal. Chemistry*, Vol. 74, pp. 1509–1518.

Hamrapurkar, P., M.G. Patil, P. Mitesh and P. Sandeep (2011). A developed and validated stability–indicating reverse phase high performance liquid chromatographic method for determination of cefdinir in the presence of its degradation products as per international conference on harmonization guidelines. *Pharm methods*, Vol. 2 , pp. 15–20.

Harayama, S.,Y.Kasai and H. Hara (2004). Microbial communities in oil–contaminated seawater, *Curr. Opin. Biotechnology*, Vol. 15, pp. 205–214.

Harms, H., D. Schlosser and L.Y. Wick (2011). Untapped potential: Exploiting fungi in bioremediation of hazardous chemicals. *Nat. Rev. Microbiology*, Vol. 9, pp. 177–192.

Harner, T., H. Kylin, T.F. Bidleman and W.M. Strachan (1999). Removal of α– and γ–hexachlorocyclohexane and enantiomers of α–hexachlorocyclohexane in the eastern Arctic Ocean, *Environ. Sci. Technology*, Vol. 33, No. 8, pp. 1157–1164.

Hassan, A.M., M. Abed El–Moteleb and A.M. Ismael (2009). Removal of lindane and malathion from wastewater by activated carbon prepared from apricot stone, *Ass. Univ. Bull. Environ. Research*, Vol. 12, pp. 1–8.

Hatha, A. A. M., K.M. Mujeeb Rahiman, K.P. Krishnan, A.V. Saramma, G. Saritha and Deepu Lal (2013). Characterisation and bioprospecting of cold adapted yeast from water samples of kongsfjord , Norwegian Artic, *Ind. J. Geo–marine Sciences*,Vol 42, No. 4, pp. 458–465.

Hill, E. F. and M.B. Camardese (1986). Lethal *dietary toxicities of environmental contaminants to coturnix, Technical report number 2, U.S. Department of Interior, Fish and Wildlife Service*, Washington, DC, pp. 6–55.

Hillis, D.M. and J.J. Bull (1993). An empirical test of bootstrapping as a method for assessing condence in phylogenetic analysis, *Syst. Biology*, Vol. 42, pp. 182–192.

Holliger, C., S. Gaspard., G. Glod., C. Heijman., W. Schumacher., R.P. Schwarzenbach and F. Vazquez (1997). Contaminated environments in the subsurface and bioremediation: Organic contaminants, *FEMS. Microbiol. Reviews*, Vol. 20, pp. 517–523.

Homem, V. and L. Santos (2011). Degradation and removal methods of antibiotics from aqueous matrices, *J. Environ. Management*, Vol. 92, pp. 2304–2347.

Hong, J.H., J. Kim., O.K. Choi., K.S. Cho and H.W. Ryu (2005). Characterization of a diesel–degrading bacterium, *Pseudomonas aeruginosa* IU5, isolated from oil contaminated soil in Korea, *World. J. Microbiol. Biotechnology*, Vol. 21, pp. 381–384.

Horsfall, M. Jnr. and A.I. Spiff (2005). Effect of temperature on the sorption of Pb^{2+} and Cd^{2+} ions from aqueous solutions by *Caladium bicolor* (Wild Cocoyam) biomass, *Electron. J. Biotechnology*, Vol. 8, No. 2, pp. 162–169.

Hu, D and J.R. Coats (2007). Aerobic degradation and photolysis of tylosin in water and soil, *Environ. Toxicol. Chemistry*, Vol. 26, pp. 884–889.

Hu, W., Y. Lu, T. Wang, W. Luo, X. Zhang, J. Geng, G. Wang, Y. Shi, W. Jiao and C. Chen (2010). Factors affecting HCH and DDT in soils around watersheds of Beijing reservoirs, China, *Environ. Geochem. Health,* Vol. 32, No. 2, pp. 85–94.

Hudel, K., F. Forge., M. Klein., H.F. Schroder and M. Dohmann (1995). Steam extraction of organically contaminated soil and residue–Process development and implementation on an industrial scale. *In*: Contaminated soils, W.J. Van den Brick., R. Bosman and F. Arendt (*eds.*), Kluwer academic publishers, Netherlands, pp. 1103–1112.

Huseyin, S.(2005). Decolorization and detoxification of textile wastewater by Ozonation and coagulation, *Dyes Pigments,* Vol. 64, pp. 217–222.

Iancu, I., A.Olmer and R.D. Strous (2007). "Caffeinism: History, clinical features, diagnosis, and treatment". *In*: Caffeine and activation theory: effects on health and behavior, Smith BD, Gupta U, Gupta BS (*ed.*) CRC Press. pp. 331–344.

Ibrahim, S.A., M.M. Salameh, S. Phetsomphou, H. Yang and C.W. Seo (2006). Application of caffeine, 1,3,7–trimethylxanthine, to control *Escherichia coli* O157: H7, *Food Chemistry*, Vol. 99, pp. 645–650.

IHPA (2009). Pesticide Storage Endangers Tens of Millions in Europe, *International HCH and Pesticides Association (IHPA) Press Release,* October, 22.

Ilori, M.O., S.A. Adebusoye and A.C. Ojo (2008). Isolation and characterization of hydrocarbon–degrading and biosurfactant–producing yeast strains obtained from a polluted lagoon water, *World. J. Microbiol. Biotechnology*, Vol. 24, pp. 2539–2545.

IPCS (1998). Hexachlorobenzene health and safety Guide. No. 107. *International Program on Chemical Safety,* World Health Organization.

Ipek, U. (2005). Removal of Ni(II) and Zn(II) from an aqueous solution by reverse osmosis, *Desalination*, Vol. 174, pp. 161–169.

Itoh, K., Y. Kitade and C. Yatome (1996). A pathway for biodegradation of an Anthroquinone Dye, C.I. Disperse Red 15, by a yeast strain *Pichia anomala, Bull. Environ. Contam. Toxicology,* Vol. 56, pp. 413–418.

Iwamoto, T. and M. Nasu (2001). Current bioremediation practice and perspective. *J. Biosci. Bioengineering*, Vol. 32, pp. 1–8.

Jacques, R.J.S., B.C. Okeke, F.M. Bento, A.S. Teixeira, M.C.R. Peralba and F.A.O. Camargo (2008). Microbial consortium bioaugmentation of a polycyclic aromatic hydrocarbons contaminated soil, *Bioresour. Technology*, Vol. 99, pp. 2637–2643.

Jadhav, J.P. and S.P.Govindwar (2006). Biotransformation of malachite green by *Saccharomyces cerevisiae* MTCC 463, *Yeast,* Vol. 23, pp. 315–323.

Jadhav, S.U., M.U. Jadhav, A.N. Kagalkar and S.P. Govindwar (2008a). Decolorization of brilliant blue G dye mediated by degradation of the microbial consortium of *Galactomyces geotrichum* and *Bacillus* sp, *J. Chin. Inst. Chem. Engineering,* Vol. 39, pp. 563–570.

Jha, V.K., M. Matsuda and M. Miyake (2008). Sorption properties of the activated carbon–zeolite composite prepared from coal fly ash for Ni^{2+}, Cu^{2+}, Cd^{2+} and Pb^{2+}, *J. Hazard. Materials,* Vol. 160, pp. 148–153.

Jogdand,S.N (2004). Environmental Biotechnology. Himalaya Publishing House, Mumbai.

Jogdand,S.N (2006). Environmental Biotechnology. Himalaya Publishing House, Mumbai. pp. 119.

Johnson, S., P.J. Maziade, L.V. McFarland, W. Trick, C. Donskey, B. Currie, D.E. Low and E.J. Goldstein (2012). Is primary prevention of *Clostridium difficile* infection possible with specific probiotics?, *Int. J. Infect. Diseases,* Vol. 16, pp. 786–792.

Johnson, W.W and M.T. Finley (1980). M. T. Handbook of Acute Toxicity of Chemicals to Fish and Aquatic Invertebrates, Resource Publication 137. *U.S. Department of Interior, Fish and Wildlife Service,* Washington, DC, pp. 6–56.

Joy, R.M. (1982). Mode of action of lindane, dieldrin and related insecticides in the central nervous system, *Neurobeh. Toxicol. Teratology,* Vol. 4, pp. 813–823.

Ju, Y.H., T.C. Chen and J.C. Liu (1997). A study on the biosorption of lindane, *Colloids Surf. B Biointerfaces,* Vol. 9, pp. 187–196.

Junker, T., R. Alexy, T. Knacker and K. Kümmerer (2006). Biodegradability of ^{14}C labelled antibiotics in a modified laboratory scale sewage treatment plant at environmentally relevant concentrations, *Environ. Sci. Technology,* Vol. 40, pp. 318–326.

Kanaly, R.A and H.G. Hur (2006). Growth of *Phanerochaete chrysosporium* on diesel fuel hydrocarbons at neutral pH, *Chemosphere,* Vol. 63, pp. 202–211.

Kato, M., K. Mizuno, A. Crozier, T. Fujimura and H. Ashihara (2000). Plant biotechnology: Caffeine synthase gene from tea leaves. *Nature,* Vol. 406, pp. 956–957.

Katsou, E., S. Malamis, M.T. Katherine, J. Haralambous and M. Loizidou (2011). Regeneration of natural zeolite polluted by lead and zinc in wastewater treatment systems. *J. Hazard. Materials,* Vol. 189, pp. 773–786.

Kawahara, H., A. Hirai, T. Minabe and H. Obata (2013). Stabilization of astaxanthin by a novel biosurfactant produced by *Rhodotorula mucilaginosa* KUGPP–1, *Biocontrol Science,* Vol. 18, No. 1, pp. 21–28.

Kelewou, H., M, Merzouki and A. Lhassani (2014). Biosorption of textile dyes Basic Yellow 2 (BY2) and Basic Green 4 (BG4) by the live yeast S.*cerevisiae*. *J. Mater. Environ. Science,* Vol. 5, pp. 633–640.

Khan, A.J., A. Sharma, K. Dinesh and D. Parmar (2013). Similarities in lindane induced alteration in cytochrome P450s and associated signaling events in peripheral blood lymphocytes and brain, *Food Chem. Toxicology*, Vol. 60C, pp. 318–327.

Khehra, M.S., H.S. Saini, D.K. Sharma, B.S. Chadha, and S.S. Chimni (2006). Biodegradation of azo dye C.I. Acid Red 88 by an anoxic–aerobic sequential bioreactor, *Dyes Pigments*, Vol. 70, pp. 1–7.

Kim, S.C. and K. Carlson (2006). Occurrence of ionophore antibiotics in water and sediments of a mixed–landscape watershed, *Water Research*, Vol. 40, pp. 2549–2560.

Kim,T.Y., J.W.Lee and S.Y. Cho (2014). Application of residual brewery yeast for adsorption removal of Reactive Orange 16 from aqueous solution, *Adv. powdered technology*. doi:10.1016/j.apt.2014.10.006.

Klaassen, E.D. (2001). Heavy metal and heavy metal antagonists, In: Hardman, J.G., Limbird, L.E., Gilman, A.G. (*eds.*), Goodman & Gilman's: The pharmacological Basis of Therapeutics, 9th ed. McGraw Hill, New York, pp. 1851–1875.

Klauson, D., J. Babkina, K. Stepanova, M. Krichevskaya and S. Preis (2010). Aqueous photocatalytic oxidation of amoxicillin, *Catal. Today*, Vol. 151, pp. 39–45.

Knop,M. (2011). Yeast cell morphology and sexual reproduction–A short overview and some considerations, *C. R. Biologies*, Vol. 334, pp. 599–606.

Knoshaug, E.P., J.A. Ahlgren and J.A. Trempy (2000). Growth Associated Exopolysaccharide Expression in *Lactococcus lactis* subspecies *cremoris* Ropy352, *J. Dairy Science*, Vol. 83, pp. 633–640.

Korngold, E., N. Belayev and L. Aronov (2003). Removal of chromates from drinking water by anion exchangers, *Sep. Purifi. Technology*, Vol. 33, pp. 179–187.

Krishnan, S., B. Roach, K. Kasinathan, P. Annamalai, T. Nooruddin and M. Gunasekaran (2012). Studies on the biodegradation of cephalosporin drugs in pharmaceutical effluent using *Pseudomonas putida* and *Pseudomonas fluorescence*, *Botany*, pp. 7–11.

Kuhn, S.P. and R.M. Pfister (1990). Accumulation of cadmium by immobilized *Zooloea ramigera* 115, *Indian. Mirobiology*, Vol. **6,** pp. 123–128.

Kumari, B., J. Singh, S. Singh and T.S. Kathpal (2005). Monitoring of butter and ghee (claried butter fat) for pesticidal contamination from cotton belt of Haryana, India, *Environ. Monit. Assessment*, Vol. 105, pp. 111–120.

Kumari, B., V.K. Madan and T.S. Kathpal (2006). Monitoring of pesticides residues in fruits, *Environ. Monit. Assessment*, Vol. 123, pp. 407–412.

Kümmerer, K (2009). Antibiotics in the aquatic environment: A review–Part I, *Chemosphere*, Vol. 75, pp. 417–434.

Kümmerer, K and A. Henninger (2003). Promoting resistance by the emission of antibiotics from hospitals and households into effluents, *Clin. Microbiol. Infection*, Vol. 9, pp. 1203–1214.

Kutty, S. N. and R. Philip (2008). Marine yeasts—a review, *Yeast*, Vol. 25, pp. 465–483.

Kvenvolden, K.A. and C.K. Cooper (2003), Natural seepage of crude oil into the marine environment, *Geo–Mar. Letters*, Vol. 23, pp. 140–146.

Kwasiewska, K. (1985). Biodegradation of crystal violet (hexamethyl–p–rosaniline chloride) by oxidative red yeasts, *Bull. Environ. Contam. Toxicology,* Vol. 34, pp. 323–330.

Lakshmi, V. and N. Das (2010). Caffeine degradation by yeasts isolated from caffeine contaminated samples, *Int. J .Sci. Nature*, Vol. 1(1), pp. 47–52.

Lakshmi, V. and N. Das (2011). Biodegradation of caffeine by *Trichosporon asahii* isolated from caffeine contaminated soil, *Int. J. Eng. Sci. Technology,* Vol. 3, pp. 7977–7986.

Lakshmi, V. and N. Das (2012). Bioremediation of caffeine–contaminated soil by immobilized yeast–A laboratory based study, *Nat. Environ. Pollut. Technology,* Vol. 11(3), pp. 385–392.

Lakshmi, V and N. Das (2013a). Removal of caffeine from industrial wastewater using *Trichosporon asahii, J. Env. Biology,* Vol 34, pp. 701–708.

Lakshmi V, D. Das and N.Das (2013b) Biodegradation of caffeine by the yeast *Trichosporon asahii* immobilized in single and hybrid matrices. *Ind. J. Chem. Technology,*Vol. 20, pp. 195–201.

Lal, R. and D.M. Saxena (1982). Accumulation, metabolism, and effects of organochlorine insecticides on microorganisms. *Microbiol. Reviews,* Vol. 46, pp. 95–127.

Lal, R., M. Dadhwal, K. Kumari, P. Sharma, A. Singh, H. Kumari, S. Jit, S.K. Gupta, A. Nigam, D. Lal, M. Verma, J. Kaur, K. Bala, and S. Jindal (2008). *Pseudomonas* sp. to *Sphingobium indicum*: A journey of microbial degradation and bioremediation of hexachlorocyclohexane. *Indian J. Microbiology,* Vol. 48, pp. 3–18.

Leahy, J.G. and R. R. Colwell (1990). Microbial degradation of hydrocarbons in the environment, *Microbiol. Reviews,* Vol. 54, pp. 305–315.

Lee, C.J.B., M.A. Fletcher., O.I. Avila., J. Callanan., S. Yunker and D.M. Munnecke (1995). Bioaugmentation for site remediation. Pro. Pap 3rd Int. *In situ* On site Bioreclam. Sym, Battelle Press, Columbus, pp.195–202.

Leffrang, U., K. Ebbert., K. Flory., U. Galla and H. Schnieder (1995). Organic waste destruction by indirect electrooxidation, *Sep. Sci. Technology,* Vol. 30, pp. 1883–1899.

Leite de Oliveira A.M., M.A. Miguel, R.S. Peixoto, A.S. Rosado, J.T. Silva and V.M. Paschoalin (2013). Microbiological, technological and therapeutic properties of kefir: A natural probiotic beverage, *Braz. J. Microbiology,* Vol. 44, pp. 341–349.

Leson, C.L., M.A. McGuigan and S.M.Bryson (1988). Caffeine overdose in an adolescent male. *J. Toxicol. Clin. Toxicology,* Vol. 26, pp. 407–415.

Li, B., Y.Yang, Y.Gan, C.D. Eaton, P. He and A.D. Jones (2000). On–line coupling of subcritical water extraction with high–performance liquid chromatography via solid–phase trapping, *J. Chromatography A,* Vol. 873, pp. 175–184.

Li, K., A. Yediler, M. Yang, S. Schulte–Hostede and M.H. Wong (2008). Ozonation of oxytetracycline and toxicological assessment of its oxidation by-products.*Chemosphere* Vol. 72, pp. 473–478.

Li, C., W. Jianga, N. Ma, Y. Zhua, X. Donga, D. Wanga, X. Menga and Y. Xua (2013). Bioaccumulation of cadmium by growing *Zygosaccharomyces rouxii* and *Saccharomyces cerevisiae, Bioresour. Technology,* doi.org/10.1016/j.biortech. 2013.12.098 (In Press).

Lissemore, L., C. Hao, P. Yang, P.K. Sibley, S. Mabury and K.R. Solomon (2006). An exposure assessment for selected pharmaceuticals within a watershed in Southern Ontario, *Chemosphere*, Vol. 64, pp. 717–729.

Liu, Q., Y.J. Li, J. Zhang, Y. Chi, X.X. Ruan, J.Y. Liu and G.R. Qian (2011). Effective removal of zinc from aqueous solution by hydrocalumite, *Chem. Eng. Journal,* Vol. 175, pp. 33–38.

Liu H, W. Liu, J. Zhang, C. Zhang, L. Ren and Y. Li (2011). Removal of cephalexin from aqueous solutions by original and Cu(II)/Fe(III) impregnated activated carbons developed from lotus stalks Kinetics and equilibrium studies. *J. Hazard.Materials*.Vol. 185, pp. 1528–1535.

Lopez, M.C., C. Cabrea, C. Gallego and M.L. Lorenzo (1994). Cadmium levels in waters of Canada Coast, *Arch. Pharm.*, Vol. 1, pp. 945–950.

Lou, M., R. Garray and J. Alda (1991). Cadmium uptake through the anion exchange in human red blood cells, *J. Physiology*, Vol. 443, pp. 123–136.

Loudon, G.M. (1984). Organic Chemistry Reading. Massachusetts, USA, p. 1196.

Lu, J.L., M.Y. Wu, X.L.Yang, Z.B. Dong, J.H. Ye, D. Borthakur, Q.L. Sun and Y.R. Liang (2010). Decaffeination of tea extracts by using poly (acrylamide–co–ethylene glycol dimethylacrylate) as adsorbent. *J. Food Engineering*, Vol. 97, pp. 555–562.

Lu,X., X.J. Zhou and T.S. Wang (2013). Mechanism of uranium(VI) uptake by *Saccharomyces cerevisiae* under environmentally relevant conditions: Batch, HRTEM, and FTIR studies, *J. Hazard. Materials*,Vol. 262, pp. 297–303.

Lucas, M.S., C. Amaral, A. Sampaio, J.A. Peres and A.A. Dias (2006). Biodegradation of the diazo dye Reactive Black 5 by a wild isolate of *Candida oleophila, Enzyme Microb. Technology,* Vol. 39, pp. 51–55.

Lyman, W.J., D.C. Nooman and P.J. Reidy (1990). Cleanup of petroleum contaminated soils at underground storage tanks, *Pollut. Technol. Rev.* No 195, Noyes Data Corp, Park Ridge, New Jersy.

Machado, M.D., E.V. Soaresa and H.M.V.M. Soares (2010). Removal of heavy metals using a brewer's yeast strain of *Saccharomyces cerevisiae*: Chemical speciation as a tool in the prediction and improving of treatment efficiency of real electroplating effluents, *J. Hazard. Materials*, Vol. 180, pp. 347–353.

MacRae, I.C., K. Raghu and T.F. Castro (1967). Persistence and biodegradation of four common isomers of benzene hexachloride in submerged soils, *J. Agri. Food Chemistry*, Vol. 15, pp. 911–914.

Madaeni, S.S. and Y. Mansourpanah (2003). COD Removal from concentrated wastewater using membranes. *Filtr. Separation*, Vol. 40, Vol. 6, pp. 40–46.

Madhavan, A., A. Srivastava, A. Kondo and V.S. Bisaria (2012). Bioconversion of lignocellulose–derived sugars to ethanol by engineered *Saccharomyces cerevisiae*, *Crit. Rev. Biotechnology*, Vol. 32, pp. 22–48.

Mahmoud, M.E (2014). Water treatment of hexavalent chromium by gelatin–impregnated yeast (GeleYst) biosorbent. *J. Environ. Management*, Vol. 147, pp. 264 –270.

Ma³achowska–Jutsz, A., J. Mrozowska., M. Kozielska and K. Miksch (1997). Aktywnoœæ enzymatyczna w glebie ska¿onej zwi¹zkami ropopochodnymi w procesie jej detoksykacji, *Biotechnol. Journal*, Vol. 36, pp. 79–91.

Malaiyandi, M., S.M. Shah and P. Lee (1982). Fate of alpha– and gamma–hexachlorocyclohexane isomers under simulated environmental conditions, *J. Environ. Sci. Health Part A*, Vol. A17, pp. 283–297.

Manickam, N., A. Bajaj, H.S. Saini and R. Shanker (2012). Surfactant mediated enhanced biodegradation of hexachlorocyclohexane (HCH) isomers by *Sphingomonas* sp. NM05, *Biodegradation*, Vol. 23, pp. 673–682.

Manonmani, H. K., Chandrashekaraiah, D. H., Reddy, N. S., Elcey, C.D. and Kunhi, A. (2000) Isolation and acclimation of a microbial consortium for improved aerobic degradation of α–hexachlorocyclohexane, *J. Agr. Food Chemistry*, Vol. 48, pp. 4341–4351.

Manzoor, Q., N. Raziya, I. Munawar, S. Rashid and M.A. Tariq (2013). Organic acids pretreatment effect on *Rosa bourbonia* phyto–biomass for removal of Pb (II) and Cu (II) from aqueous media, *Bioresour. Technology*,Vol. 132, pp. 446–452.

Martinelli, E.E., K. Schultz and G.A. Mansoori (1991). Supercritical fluid extraction/ retrograde condensation (SFE/RC) with applications in biotechnology, *In*: Supercritical Fluid Technology, Bruno, T.J., Ely, J.F. (*ed.*), CRC Press, Boca Raton, FL, pp. 451–478.

Martínez, J.L (2009). Environmental pollution by antibiotics and by antibiotic resistance determinants. *Environ. Pollution*, Vol. 157, pp. 2893–2902.

Martini, A (1992). Biodiversity and conservation of yeasts, *Biodivers. Conservation*, Vol. 1, pp. 324–333.

Mathur, H.B., S. Jhonson, B. Singh, R. Mishra and A. Kumar (2003). Bottled water has pesticide residue. *Down Earth*, Vol. 11, pp. 27–33. http:// www.cseindia.org/html/lab/bottled water result.htm.

Mazzafera, P. (2002). Degradation of caffeine by microorganisms and potential use of decaffeinated coffee husk and pulp in animal feeding, *Sci. Agricola.*, Vol. 59, pp. 815–821.

McErlean, C., R. Marchant and I.M. Banat (2006). An evaluation of soil colonisation potential of selected fungi and their production of ligninolytic enzymes for use in soil bioremediation applications, *Antonie van Leeuwenhoek*, Vol. 90, pp. 147–158.

McFarland, L.V. (2010). Systematic review and meta analysis of *Saccharomyces boulardii* in adult patients, *World. J. Gastroenterology*, Vol. 16, pp. 2202–2222.

Méndez–Díaz, J.D., G. Prados–Joya, J. Rivera–Utrilla, R. Leyva–Ramos, M. Sánchez– Minh, T.B., H.W. Leung, I.H. Loi, W.H. Chan, M.K. So, J.Q. Mao, D. Choi, J.C.W. Lam, G. Zheng, M. Martin, J.H.W. Lee, P.K.S. Lam and B.J. Richardson (2009). Antibiotics in the Hong Kong metropolitan area: Ubiquitous distribution and fate in Victoria Harbour, *Mar. Pollut. Bulletin,* Vol. 58, pp. 1052–1062.

Mertens, B., C. Blothe, K. Windey, W. De Windt and W. Verstraete (2007). Biocatalytic dechlorination of lindane by nano–scale particles of Pd(0) deposited on *Shewanella oneidensis, Chemosphere,* Vol. 66, No. 1, pp. 99–105.

Meyer, L., J. Caston and H.R. Lieberman (2004). Stress alters caffeine actions on investigatory behavior and behavioural inhibition in the mouse, *Behav. Brain. Research,* Vol. 149, pp. 87–93.

MFE (2011).Stockholm Convention, *Ministry for Environment,* http://www.mfe.govt.nz/laws/meas/stockholm.html.

Middelhoven, W. J. and A. Lommen (1984). Degradation of caffeine by *Pseudomonas putida* C3024, *Antonie van Leeuwenhoek,* Vol. 50, pp. 298–300.

Miranda, R.C., C.S. Souza., E.B. Gomes., R.B. Lovaglio., C.E. Lopes and M.F. Sousa (2007). Biodegradation of diesel oil by yeasts isolated from the vicinity of Suape Port in the State of Pernambuco *Brazil, Braz. Arch. Biol. Technology,* Vol. 50, pp. 147–152.

Minh, T.B., H.W. Leung, I.H. Loi, W.H. Chan, M.K. So, J.Q. Mao, D. Choi, J.C.W. Lam, G. Zheng, M. Martin, *et al.* (2009). Antibiotics in the Hong Kong metropolitan area: Ubiquitous distribution and fate in Victoria Harbour, *Mar. Pollut. Bulletin,* Vol. 58, pp. 1052–1062.

Mohamed, R.S. and G.A. Mansoori (2002). The Use of Supercritical Fluid Extraction Technology in Food Processing, Featured Article–Food Technology Magazine, The World Markets Research Center, London, UK.

Molleron, H.J. (1994). Thermal treatment of petroleum contaminated soils: waste tracking and performance analysis in an off–site facility. *Pro. Therm. Treat. Radioact., Haz. Chem., Mixed, Munitions, Pharm. Wastes.* University of California, Irvine, pp. 47–53.

Mompelat, S., B. LeBot and O. Thomas (2009). Occurrence and fate of pharmaceutical products and by–products from resource to drinking water, *Environ. International,* Vol. 35, pp. 803–814.

Monem, M.A., E. Zeftawy and C.N. Mulligan (2011). Use of rhamnolipid to remove heavy metals from wastewater by micellar–enhanced ultrafiltration (MEUF), *Sep. Purif. Technology,* Vol. 77, pp. 120–127.

Moore, M.T., S.L. Greenway and J.L. Farris (2008). Assessing caffeine as an emerging environmental concern using conventional approaches, *Arch. Environ. Contam. Toxicology,* Vol. 54, pp. 31–35.

Morita, T., M. Konishi, T. Fukuoka, T. Imura and D. Kitamoto (2006). Discovery of *Pseudozyma rugulosa* NBRC 10877 as a novel producer of the glycolipid biosurfactants, mannosylerythritol lipids, based on rDNA sequence, *Appl. Microbiol. Biotechnology,* Vol. 73, pp. 305–313.

Muhan, B.S. and P. Agrawal (2006). Is phytoremediation the solution for arsenic contamination of groundwater in India and Bangladesh? *Curr. Sci. India*, Vol. 90 (4), pp. 476–477.

Munk, Z.M. and A. Nantel (1977). Acute lindane poisoning with development of muscle necrosis, *Can. Med. Assoc. Journal*, Vol. 117, pp. 1050–1054.

Murthy, H.M. and H.K. Manonmani (2007). Aerobic degradation of technical hexachlorocyclohexane by a defined microbial consortium, *J. Hazard. Materials*, Vol. 149, No. 1, pp. 18–25.

Muruganandham, M. and M. Swaminathan (2004). Photochemical oxidation of reactive azo dye with UV–H_2O_2 process, *Dyes Pigments*, Vol. 62, pp. 269–275.

Nagpal V, M.C. Srinivasan and K.M. Paknikar (2008). Biodegradation of γ–hexachlorocyclohexane (lindane) by a non–white rot fungus *Conidiobolus* 03–1–56 isolated from litter, *Indian J. Microbiology*, Vol. 48, pp. 134–141.

Nagpal, V. and K.M. Paknikar (2006). Integrated biological approach for enhanced degradation of lindane, *Indian J. Biotechnology*, Vol. 5, pp. 400–406.

Namasivayam, C. and S. Senthilkumar (1998). Removal of arsenic from solid waste. Adsorption rates and equilibrium studies, *Ind. Eng. Chem. Research*, Vol. 37, pp. 4816–4822.

Nandan, B.S. and P.J. Nimila (2012). Lindane toxicity: Histopathological, behavioural and biochemical changes in *Etroplus maculatus* (Bloch, 1795), *Mar. Environ. Research*, Vol. 76, pp. 63–70.

Nguyen–nhu, N.T. and B. Knoops (2002). Alkyl hydroperoxide reductase 1 protects *Saccharomyces cerevisiae* against metal ion toxicity and glutathione depletion, *Toxicol Lett*ers, Vol. 135, pp. 219–228.

Nicola, D., R. Hall, N. Bollag, T. Thermogiannis and G. Walker (2009). Zinc accumulation and utilization by wine yeasts, *Int. J. Wine Research*, Vol. 1, pp. 85–94.

Nies, D.H.(1999). Microbial heavy metal resistance, *Appl. Microbiol. Biotechnol*ogy, Vol. 51, pp. 730–750.

Nolan, K., J. Kamrath and J. Levitt (2012). Lindane toxicity: A comprehensive review of the medical literature, *Pediatr. Dermatology*, Vol. 29, pp. 141–146.

NTP (1998). National Toxicology Program. NTP Chemical Repository (Radian Corporation, August 29, 1991). Lindane. Internet Site. http://ntp-db.niehs.gov/NTP_Reports/NTP_Chem_H&S/NTP_Chem5/Radian58–89–9.txt.

Nwuha, N. (2000). Novel studies on membrane extraction of bioactive components of green tea in organic solvents: part I, *J. Food Engineering*, Vol. 44, pp. 233–238.

Ogunseitan, O.A. (2002). Caffeine–inducible enzyme activity in *Pseudomonas putida* ATCC 700097, *World J. Microbiol. Biotechnology*, Vol. 18, pp. 423–428.

Okoh, A.I. (2006). Biodegradation alternative in the cleanup of petroleum hydrocarbon pollutants, *Biotechnol. Mol. Biol. Reviews*, Vol. 2, pp. 38–50.

Ollikka, P., K.Alhonmaki, V.M. Leppanen, T.Glumoff, T. Raijola and I. Suominen (1993). Decolorization of azo, triphenylmethane, heterocyclic, and polymeric dyes by lignin peroxidase isoenzymes from *Phanerocheate chrysosporium*, *Appl. Environ. Microbiology*, Vol. 59, pp. 4010–4016.

Ortiz, J.B., M.L. González de Canales and C. Sarasquete (2001). The impact of persistent organochlorine contaminant (lindane, γ–HCH): histopathological alterations in fish tissues, *Ecotox. Environ. Restoration*, Vol. 4, No. 1, pp. 45–52.

Oyeka, C.A. and L.O. Ugwu (2002). Fungal flora of human toe webs, *Mycoses*, Vol. 45, pp. 488– 491.

Ozer, A. and D. Ozer (2003). Comparative study of the biosorption of Pb (II), Ni (II) and Cr (VI) ions onto *S. cerevisiae*: Determination of biosorption heats, *J. Hazard. Mate*rials,Vol. B.100, pp. 219–229.

Pajot, H.F., L.I.C. De Figueroa and J.I. Fariña (2007). Dye–decolorizing activity in isolated yeasts from the ecoregion of Las Yungas (Tucumán, Argentina), *Enzyme Microb. Technology*, Vol. 40, pp. 1503–1511.

Pall, B. and M. Sharan (2002). Enhanced photocatalytic activity of highly porous ZnO thin films prepared by sol–gel process, *Mater. Chem. Physics*, Vol. 76, pp. 82–87.

Palominos, R., J. Freer, M.A. Mondaca and H.D. Mansilla (2008). Evidence for hole participation during photocatalytic oxidation of the antibiotic flumequine, *J. Photochem. Photobiology A*, Vol. 193, pp. 139–145.

Pan, X., G. Niu and H.Liu (2003). Microwave–assisted extraction of tea polyphenols and tea caffeine from green tea leaves, *Chem. Eng. and Pro*cessing, Vol. 42, pp. 129. —

Pandey, A., C.R. Soccol, P. Nigam, D. Brand, R. Mohan and S. Roussos (2000). Biotechnological potential of coffee pulp and coffee husk for bioprocesses, *Biochem. Eng. Journal*, Vol. 6, pp. 153–162.

Pandit, G.G. and S.K. Sahu (2002). Assessment of risk to public health posed by persistent organochlorinde pesticide residue in milk and milk products in Mumbai, *India. J Environ Monitoring*, Vol. 4, pp. 182–185.

Park, H.S., H.K. Choi, S.J. Lee, K.W. Park, S.G. Choi and K.H. Kim (2007). Effect of mass transfer on the removal of caffeine from green tea by supercritical carbon dioxide, *J. Supercritical Fluids*, Vol. 42, pp. 205–212.

Park, J.K., J.W. Lee and J.Y. Jung (2003). Cadmium uptake capacity of two strains of *Saccharomyces cerevisiae* cells, *Enzyme Microb. Techno*logy, Vol. 33, pp. 371–378.

Park, S. and W.Y. Jung (2001). Removal of chromium by activated carbon fibers plated with copper metal, *Carbon Science*, Vol. 2 (11), pp. 15–21.

Parshetti, G.K, S.D. Kalme, S.S. Gomare and S.P. Govindwar (2007). Biodegradation of Reactive Blue–25 by *Aspergillus ochraceus* NCIM–1146, *Bioresour. Technology*, Vol. **98,** pp. 3638–3642.

Pazou, E.Y.A., M. Boko, C.A.M. van Gestel, H. Ahissou, P. Laleye, S. Akpona, B. van Hattum, K. Swart and N.I. van Straalen (2006). Organochlorine and organophosphorous pesticide residues in the Oueme river catchment in the Republic of Benin, *Environ International*, Vol. 32, pp. 616–623.

Perez–Corona, T., Y. Madrid and C. Camara (1997). Evaluation of selective uptake of selenium (Se (IV) and Se (VI)) and antimony (Sb (III) and Sb (V)) species by baker's yeast cells *(Saccharomyces cerevisiae)*, *Anal. Chim. Acta.*, Vol. 345, pp. 249–255.

Perry, J.J. (1984). Microbial metabolism of cyclic alkanes, in Petroleum Microbiology, *In*: R.M. Atlas (*ed.*), Macmillan, New York, NY, USA. pp. 61–98.

Pincheira, J., J.F. L´opez–S´aez, P. Carrera, M.H. Navarreteb and C.D.L. Torre (2003). Effect of caffeine on *in vivo* processing of alkylated bases in proliferating plant cells, *Cell. Biol. Inernational*, Vol. 27, pp. 837–843.

Podgorskii, V.S., T.P. Kasatkina and O.G. Lozovaia (2004). Yeasts–biosorbents of heavy metals, *Mikrobiol Z*, Vol. 66, pp. 91–103.

Pointing, S.B. and L.L.P. Vrijmoed (2000).Decolorization of azo and triphenylmethane dyes by *Pycnoporus sanguineous* producing laccase as the sole phenol oxidase, *World J. Microbiol. Biotechnology*, Vol. 16, pp. 317–318.

Pollack, K., K. Balazs and O. Ogunseitan (2009). Proteomic assessment of caffeine effects on coral symbionts, *Environ. Sci. Technology*, Vol. 43, pp. 2085–2091.

Polman, J.K and C.R. Breckenridge (1996). Biomass mediated binding and recovery of textile dyes from waste effluents, *Text. Chem. Color*, Vol. 28, pp. 31–35.

Polti, M.A., J.D. Aparicio, C.S. Benimeli, and M.J. Amoroso (2014). Simultaneous bioremediation of Cr (VI) and lindane in soil by actinobacteria, *Int. Biodeter. Biodegradation*, Vol. 88, pp. 48–55.

Poulios, I., D. Makri and X. Prohaska (1999). Photocatalytic treatment of olive milling wastewater: Oxidation of protocatechuic acid, *Global Nest Int. Journal*, Vol. 1, pp. 55–62.

Prabua, C.S. and A.J. Thatheyus (2007). Biodegradation of acrylamide employing free and immobilized cells of *Pseudomonas aeruginosa*, *Int. Biodeterior. Biodegradation*, Vol. 60, pp. 69–73.

Prakash, O., M. Suar, V. Raina, C. Dogra, R. Pal and R. Lal (2004). Residues of hexachlorocyclohexane isomers in soil and water samples from Delhi and adjoining area, *Curr. Science*, Vol. 87, pp. 73–77.

Prakash, S., G.S. Tandon, T.D. Seth and P.C. Joshi (1994). The role of reactive oxygen species in the degradation of lindane and DDT, *Biochem. Biophys. Res. Communication*, Vol. 199, pp. 1284–1288.

Prasad, K.S., V. Subramanian and J. Paul (2011). Biosorption of As(III) ion on *Rhodococcus sp.* WB–12: Biomass characterization and kinetic studies, *Sep. Sci. Technology*, Vol. 46(16) , pp. 2517–2525.

Puranik, P.R. and K.M. Paknikar (1997). Biosorption of lead and zinc from solutions using *Streptoverticillium cinnamoneum* waste biomass, *J. Biotechnology*, Vol. 55, pp. 113–124.

Putra, E.K., R. Pranowo, J. Sunarso, N. Indraswati and S.Ismadji (2009). Performance of activated carbon and bentonite for adsorption of amoxicillin from wastewater: mechanism, isotherms and kinetics, *Water Research*, Vol. 43, pp. 2419–2430.

Qaiser, S., R.A. Saleemi and M. Umar (2009). Biosorption of lead (II) and chromium (VI) on groundnut hull: Equilibrium, kinetics and thermodynamics study, *J. Biotechnology*, Vol. 12, pp. 1–17.

Rahmanian, B., M. Pakizeh and A. Maskooki (2010). Micellar–enhanced ultrafiltration of zinc in synthetic wastewater using spiral–wound membrane, *J. Hazard. Materials*, Vol. 184, pp. 261–267.

Rajendran, P. and P. Gunasekaran (2006). Microbial Bioremediation. MJP Publishers, Chennai, pp. 67–69.

Ram, N.M., D.H. Bass., R. Falotico and M. Leahy (1993). A decision framework for selecting remediation technologies at hydrocarbon contaminated sites, *J. Soil. Contamination*, Vol. 2, No. 2, pp. 167–189.

Ramalho, P.A., H. Scholxer, M.H.Cardoso, T.A. Ramalho and M. Oliveira–Campos (2002). Improved conditions for the aerobic reductive decolorisation of azo dyes by *Candida zeylanoides*, *Enzyme Microb. Technology*, Vol. 31, pp. 848–854.

Ramalho, P.A. (2004). Degradation of dyes with microorganisms studies with ascomycete yeasts. Ph.D Thesis submitted to the University of Minho.

Rama Krishna, K. and L. Philip (2008). Adsorption and desorption characteristics of lindane, carbofuran and methyl parathion on various Indian soils, *J. Hazard. Materials*, Vol. 160, pp. 559–567.

Ramanaviciene, A., V. Mostovojus, I. Bachmotova and A. Ramanavicius (2003). Anti–bacterial effect of caffeine on *Escherichia coli* and *Pseudomonas fluorescens*, *Acta Medica Lituanica*, Vol. 10, pp. 185–188.

Ramarethinam, S. and N. Rajalakshmi (2004). Caffeine in tea plants [*Camellia sinensis* (L.) O. Kuntze]: *in situ* lowering by *Bacillus licheniformis* (Weigmann) Chester, *Indian J. Exp. Biology*, Vol. 42, pp. 575–580.

Rao, R.S., R.S. Prakasham, K.K. Prasad, S. Rajesham, P.N. Sarma and L. Rao (2004). Xylitol production by *Candida* sp.: parameter optimization using Taguchi approach, *Process Biochemistry*, Vol. 39, pp. 951–956.

Ratola, N., C. Botelho and A. Alves (2003). The use of pine bark as a natural adsorbent for persistent organic pollutants–study of lindane and heptachlor adsorption, *J. Chem. Technol. Biotechnology*, Vol. 78, pp. 347–351.

Refaat, A.A., N.K. Attia., H.A. Sibak., H.T. Sheltawy and G.I. Diwani (2008). Production, optimization and quality assessment of biodiesel from waste vegetable oil, *Int. J. Environ. Sci. Technologyy*, Vol. 5, pp. 75–82.

Reiser, J.,V. Glumoff, U.O. Ochsner and A. Fiechter (1994). Molecular analysis of the *Trichosporon cutaneum* DSM 70698 *argA* gene and its use for DNA mediated transformations, *J. Bacteriol*, Vol. 176, pp. 3021–3032.

Rivas, F.J., O. Gimeno and T. Borallho (2012). Aqueous pharmaceutical compounds removal by potassium monopersulfate. Uncatalyzed and catalyzed semicontinuous experiments, *Chem. Eng. Journal*, Vol. 192, pp. 326–333.

Rizvi, S.J.H., V.Rizvi, D.Mukerjee and S.N. Mathur (1987).1,3,7–trimethylxanthine—an allelochemical from seeds of *Coffea arabica*: some aspects of its mode of action as a natural herbicide, *Plant Soil*, Vol. 98, pp. 81–91.

Robinson, T., B. Chandran, G.–S. Naidu and P. Nigam (2002). Studies on the removal of dyes from a synthetic textile effluent using barley husk in static–

bath and in a continuous flow packed–bed reactor, *Bioresour. Technology*, Vol. 85, pp. 43–49.

Robinson, T., G. McMullan, R. Marchant and P. Nigam (2001). Remediation of dyes in textile effluent: a critical review on current treatment technologies, *Bioresour. Technology*, Vol. 77, pp. 247–255.

Robledo–Ortíz J R, D.E. Ramírez–Arreola, A.A.Pérez–Fonseca, C. Gómez, O. González–Reynoso , J. Ramos–Quirarte and R. González–Núñez (2011), Benzene, toluene, and o–xylene *degradation* by free and *immobilized P. putida* F1 of post consumer agave– fibre/polymer foamed composites, *Int. Biodeter. Biodegration*, Vol. 65 , No. 3, pp. 539–546.

Roepcke, C.B.S., L.P.S. Vandenberghe and C.R. Soccola (2011). Optimized production of *Pichia guilliermondii* biomass with zinc accumulation by fermentation, *Ani. Feed Sci. Technology*, Vol. 163, pp. 33–42.

Rozas, O., D. Contreras, M.A. Mondaca, M. Pérez–Moya and H.D. Mansilla (2010). Experimental design of Fenton and photo–Fenton reactions for the treatment of ampicillin solutions, *J. Hazard. Materials*, Vol. 177, pp. 1025–1030.

Ruberto, L., S.C. Vazquez and W.P. Mac Cormack (2003). Effectiveness of the natural bacterial flora, biostimulation and bioaugmentation on the bioremediation of a hydrocarbon contaminated Antarctic soil, *Int. Biodeter. Biodegradation*, Vol. 52, pp. 115–125.

Sadettin, S. and G. Donmez (2006). Bioaccumulation of reactive dyes by thermophilic cyanobacteria, *Process Biochemistry*, Vol. 41, pp. 836–841.

Saifuddin, N.M. and P. Kumaran (2005). Removal of heavy metal from industrial wastewater using chitosan coated oil palm shell charcoal, *Electron. J. Biotechnology*, Vol. 8, pp. 43–53.

Salam, J.A. and N. Das (2012).Remediation of lindane from environment–An overview. *Intl. J. Adv. Biol. Research*, Vol. 2, pp. 9–15.

Salam, J.A., V. Lakshmi, D. Das and N. Das (2013). Biodegradation of lindane using a novel yeast strain *Rhodotorula* sp. VITJzN03 isolated from agricultural soil, *World J. Microbiol. Biotechnology*, Vol. 29, pp. 475–487.

Salam, J.A. and N. Das (2013a). Enhanced biodegradation of lindane using oil–in–water bio–microemulsion stabilized by biosurfactant produced by a new yeast strain *Pseudozyma* VITJzN01, *J. Microbiol. Biotechnology*, Vol. 23, pp. 1598–1609.

Salam, J.A. and N. Das (2013b). Induced biosurfactant production and degradation of lindane by soil basidiomycetes yeast, *Rhodotorula* sp. VITJzN03. *Res. J. Pharma. Biol. Chem. Sciences*, Vol. 4, pp. 644–670.

Salam, J.A. and N. Das (2013c). Biosorptive removal of lindane using pretreated dried yeast *Cintractia sorghi* VITJzN02–equilibrium and kinetic studies, *Int. J. Pharma. Pharma. Sciences*, Vol. 5, pp. 98–99.

Salam, J.A. and N. Das (2014). Lindane degradation by *Candida* VITJzN04, a newly isolated yeast strain from contaminated soil – Kinetic study, enzyme analysis and biodegradation pathway, *World J. Microbiol. Biotechnology*, Vol. 30, No. 4, pp. 1301–1313.

Salam,J.A. and N.Das (2015) Degradation of lindane by a novel embedded bio–nano hybrid system in aqueous environment. *Appl. Microbiol. Biotechnology* Vol. 99, pp. 2351–2360.

Sandhu, D. K., M.K. Waraich and Waraich (1985). Yeasts associated with pollinating bees and flower nectar, *Microbial Ecology,* Vol. 11, pp. 51–58.

Sandlie, I. and K. Kleppe (1980). Mechanism of inhibition of thymidine kinase from *Escherichia coli* by caffeine, *FEBS Letters,* Vol. 110, pp. 223–226.

Sandlie, I. and K. Kleppe (1982). Effect of caffeine on nucleotide pools in *Escherichia coli, Chem. Biol.* Inter, Vol. 40, No. 2, pp. 141–148.

Sani, R.K. and U.C. Banerjee (1999). Decolourization of triphenylmethane dyes and textile and dye–stuff effluent by *Kurthia* sp, *Enzyme Microb. Technology,* Vol. 24, pp. 433–437.

Santana–Casiano, J.M. and M. Gonzalez–Davila (1992). Characterization of the sorption and desorption of lindane to chitin in seawater using reversible and resistant components, *Environ. Sci. Technology,* Vol. 26, pp. 90–95.

Sarasa, J., M.P.Roche, M.P. Ormad, E. Gimeno, A. Puig and J.L. Ovelleiro (1998). Treatment of wastewater resulting from dye manufacturing with ozone and chemical coagulation, *Water Research,* Vol. 9, pp. 2721–2727.

Saratale, R.G., G.D. Saratale, H.S., Chang and S.P. Govindwar (2009). Decolorization and biodegradation of textile dye Navy blue HER by *Trichosporon beigelii* NCIM–3326, *J.Hazard. Materials,* Vol. 166, pp. 1421–1428.

Saravanan, P., S.R. Prabagaran, Y.V. Nancharaiah, M. Krishnaveni, V.P. Venugopalan, V. P. and S. Jayachandran (2008). Isolation and characterization of *Pseudoalteromonas ruthenica* (SBT033), an EPS–producing biofilm bacterium from the seawater intake point of a tropical power station, *World J. Microbiol. Biotechnology,* Vol. 24, pp. 509–515.

Savel, S. (2003). Bioremediation: Clean–up biotechnologies for soils and aquifers, *In*: Environmental Biotechnology and Cleaner Bioprocesses (*ed.*) Olguin, E.J. Sanchez,G and Hernandez, E. Taylor and Francis Limited, Philadelphia, pp. 155–166.

Schumann, K. (1990). The toxicological estimation of the heavy metal content (Cd, Hg, Pb) in food for infants and small children, *Z. Emahrungswiss.*, Vol. 29, pp. 54–73.

Seiler R. L., S.D. Zaugg, J.M. Thomas and D.L.Howcroft (2005). Caffeine and pharmaceuticals as indicators of waste water contamination in wells, *Ground Water,* Vol. 37, pp. 405–410.

Seki, H., A. Suzuki and H. Maruyama (2005). Biosorption of Cr (VI) and Arsenic (V) onto methylated yeast biomass, *J. Colloid Interface Science,* Vol. 281, pp. 261–266.

Selvi.A, J.A, Salam and N. Das (2014) Biodegradation of cefdinir by a novel yeast strain, *Ustilago sp.* SMN03 isolated from pharmaceutical wastewater, *World J.Microbiol. Biotechnology* Vol. 30, pp. 2839–2850.

Selvi, A. and N. Das (2014a) Isolation, screening and identification of cefdinir degrading yeast for the treatment of pharmaceutical waste water. *Int. J. Pharm. Pharm. science,* Vol. 6, No. 8, pp. 382–386.

Selvi, A. and N. Das (2014b) Role of yeast isolates for degradation of third generation cephalosporin antibiotics: Cefotaxime and cefoperazone. *Int. J. Pharm. Pharm. science*, Vol. 6, No. 11, pp. 477–481.

Selvi, A. And N. Das (2015) Biosorptive removal of cefdinir from aqueous solution using dead yeast, *Candida sp.* SMNO4 biomass–Equilibrium, kinetic and thermodynamic studies, *Der Pharmacia letters* Vol. 7 , Vol. 4 , pp. 74–81.

Sharma, D.K., H.S. Saini, M. Singh, S.S. Chimni and B.S. Chadha (2004). Isoloation and characterization of microorganisms capable of decolorizing various triphenylmethane dyes, *J. Basic Microbiology*, Vol. 44, pp. 59–65.

Sharma, S., P. Singh, M. Raj, B.S. Chadha and H.S. Saini (2009). Aqueous phase partitioning of hexachlorocyclohexane (HCH) isomers by biosurfactant produced by *Pseudomonas aeruginosa* WH–2, *J. Hazard. Material*, Vol. 171, pp. 1178–1182.

Shek, T.H., A. Ma, V.K.C. Lee and G.Mc. Kay (2009). Kinetics of zinc ions removal from effluents using ion exchange resin, *Chem. Eng. Journal*, Vol. 146, pp. 63–70.

Shenker, M., D. Harush, J. Ben–Ari and B. Chefetz (2011). Uptake of carbamazepine by cucumber plants: A case study related to irrigation with reclaimed wastewater, *Chemosphere*, Vol. 82, pp. 905–910.

Shroff, K.A. and V.K. Vaidya (2012). Effect of pre–treatments on the biosorption of chromium (VI) ions by dead biomass of *Rhizopus arrhizus*, *J. Chem. Technol. Biotechnology*, Vol. 87(2), pp. 294–304.

Shumilla, A.J., J.R. Broderick, Y. Wang and A. Barchowsky (1999). Chromium Cr (VI) inhibits the transcriptional activity of nuclear factor–B by decreasing the interaction of p^{65} with cAMP–responsive element–binding protein, *J. Biol. Chem*istry, Vol. 274, No. 52, pp. 36207–36212.

Siddique, T., B.C. Okeke, M. Arshad and W.T. Frankenberger (2002). Temperature and pH effects on biodegradation of hexachlorocyclohexane isomers in water and a soil slurry, *J. Agr. Food Chemistry*, Vol. 50, pp. 5070–5076.

Silar, P., J. Dairou, A. Cocaign, F. Busi, F. Rodrigues–Lima and J.M. Dupret (2011). Fungi as a promising tool for bioremediation of soils contaminated with aromatic amines, a major class of pollutants, *Nat. Rev. Microbiology*, Vol. 9(6), pp. 477.

Silva, D. de. P., D. De. L. Cavalcanti, E.J.V. de. Melo, P.N.F. dos. Santos, *et. al.* (2015). Bioremoval of diesel oil through a microbial consortium isolated from a polluted environment. *Int. Biodeterior.Biodegradation*, Vol. 97, pp. 85–89.

Singh, D.N. and A.K. Tripathi (2013). Coal induced production of a rhamnolipid biosurfactant by *Pseudomonas stutzeri*, isolated from the formation water of Jharia coalbed. *Bioresour. Technology*,Vol. 128, pp. 215–221.

Singh, K.P., A. Malik and S. Sinha (2007). Persistent organochlorine pesticide residues in soil and surface water of northern Indo Gangetic alluvial plains, *Environ. Monit. Assessment*, Vol. 125, pp. 147–155.

Sláviková, E. and R. Vadkertiová (2003). The diversity of yeasts in the agricultural soil, *J. Basic. Microbiology*, Vol. 43, pp. 430–436.

Slokar, Y.M. and A.M. Le Marechal (1998). Methods of decolorization of textile wastewaters, *Dyes and Pigments*, Vol. 37, pp. 335–356.

Smyth, D.A. (1992). Effect of methylxanthine treatment on rice seedling growth, *J. Plant Growth Regulators*, Vol. 11, pp. 125–128.

Soares, E.V. and H.M.V.M. Soares (2012). Bioremediation of industrial effluents containing heavy metals using brewing cells of *Saccharomyces cerevisiae* as a green technology: A review, *Environ. Sci. Pollut. Research*, Vol. 19, pp. 1066–1083.

Soares, E.V., G.D. Coninck, F. Duarte and H.M.V.M. Soares, (2002). Use of *Saccharomyces cerevisiae* for Cu^{2+} removal from solution: The advantages of using a flocculent strain, *Biotechnol. Let*ters, Vol. 24, pp. 663–666.

Song, C., L. Sheng and X. Zhang (2012). Preparation and characterization of a thermostable enzyme (Mn–SOD) immobilized on supermagnetic nanoparticles, *Appl. Microbiol. Biotechnology*, Vol. 96, No. 1, pp. 123–132.

Sotelo, J.L.,G. Ovejero, J.A. Delgado and I. Martinez (2002). Adsorption of lindane from water onto GAC: effect of carbon loading on kinetic behaviour, *Chem. Eng. Journal*, Vol. 87, pp. 111–120.

Spiff, A.I. and A.A. Uwakwe (2003). Activation of human erythrocyte glutathione–s–transferase (EC.2.5.1.18) by caffeine (1,3,7–trimethylxanthine), *J. Appl. Sci. Environ. Management*, Vol. 7, pp. 45–48.

Stanislaw, L. and G. Monika (1999).Optimization of oxidants dose for combined chemical and biological treatment of textile wastewater, *Water Research*. Vol. 33, pp. 2511–2516.

Suazo–Madrid, A., L. Morales–Barrera, E. Aranda–García and E. Cristiani–Urbina (2011). Nickel(II) biosorption by *Rhodotorula glutinis*, *J. Ind. Microbiol. Biotechnology*, Vol. 38, pp. 51–64.

Suh, S.O., J.V. McHugh, D.D. Pollock and M. Blackwell (2005). The beetle gut: A hyperdiverse source of novel yeasts, *Mycol. Research*, Vol. 109, pp. 261–265.

Sun, Q.L., H. Shu, J.H. Ye, J.L. Lu, X.Q. Zheng and Y.R. Liang (2010). Decaffeination of green tea by supercritical carbon dioxide, *J. Med. Plants Research*, Vol. 4, pp. 1161–1168.

Sundararaman, S. and R.Saravanane (2010) Effect of loading rate and HRT on the removal of cephalosporin and their intermediates during the operation of a membrane bioreactor treating pharmaceutical wastewater. Water Sci Technology,Vol. 61, pp. 1907–1914.

Sunder, R.R.C.V., R. Shreenivas and V. Singh (1988). Disseminated intravascular coagulation in a case of fatal lindane poisoning, *Vet. Hum. Toxicology*, Vol. 30, pp. 132–134.

Talebi, K., A. Kavousi and Q. Sabahi (2008). Impacts of pesticides on arthropod biological control agents, *Pest Technology*, Vol. 2, pp. 87–97.

Tamura, K., D. Peterson, N. Peterson, G. Stecher, M. Nei and S. Kumar (2011). MEGA5: molecular evolutionary genetics analysis using maximum likelihood, evolutionary distance, and maximum parsimony methods, *Mol. Biol. Evolution*, Vol. 28, pp. 2731–2739.

Tan, C.S., H.C. Lien, S.R. Lin, H.L. Cheng and K.J. Chao (2003). Separation of supercritical carbon dioxide and caffeine with mesoporous silica and microporous silicalite membranes, *J. Super crit. Fluids*, Vol. 26, pp. 55–62.

Tan, L., S. Ning, X. Zhang and S. Shi (2013) Aerobic decolorization and degradation of azo dyes by growing cells of a newly isolated yeast *Candida tropicalis* TL-F1.*Bioresour. Technology*, Vol. 138, pp. 307–313.

Tanaka, E., A. Ishikawa, Y. Yamamoto, A. Osada, K. Tsuji, K. Fukao and Y. Iwasaki (1993). Comparison of hepatic drug oxidising activity after simultaneous administration of two probe drugs caffeine and trimethedone to human subjects, *Pharmacol. Toxicology*, Vol. 72, pp. 31–33.

Tarangini, K., A. Kumar, G.R. Satpathy and V.K. Sangal (2009). Statistical optimization of process parameters for Cr (VI) biosorption onto mixed cultures of *Pseudomonas aeruginosa* and *Bacillus subtilis*, *Clean– Soil, Air, Water*, Vol. 37, pp. 319–327.

Teberikler, L., S.S. Koseonglu and A. Akgerman (2001). Deoling of crude lecithin using supercritical carbon dioxide in the presence of co–solvents, *J. Food Science*, Vol. 66, pp. 850–856.

Theodoridis, G. and P.Manesiotis (2002). Selective solid–phase extraction sorbent for caffeine made by molecular imprinting, *J. Chromatography A*, Vol. 948, pp. 163–169.

Topp, E., J. G. Hendel, Z. Lu, and R. Chapman (2006). Biodegradation of caffeine in agricultural soils, *Can. J. Soil Science*, Vol. 86, pp. 533–544.

Trovó, A.G., S.A.S. Melo and R.F.P. Nogueira (2008). Photodegradation of the pharmaceuticals amoxicillin, bezafibrate and paracetamol by the photo–Fenton process application to sewage treatment plant effluent, *J. Photochem. Photobiology A*, Vol. 198, pp. 215–220.

Tsezos, M. and J.P. Bell (1989). Comparison of the biosorption and desorption of hazardous organic pollutants by live and dead biomass, *Water Research*, Vol. 23, pp. 561–568.

Udayasankar, K., B. Manaohar and A. Chokkalingam (1986). A note on super-critical carbondioxide decaffeination of coffee, *J. Food Sci. Technology*, Vol. 23, pp. 326–328.

Udayasankar, K., C.V. Raghavan, K.L. Rao, S. Kuppuswamy and P.K. Ramanathan (1983). Studies on extraction of caffeine from coffee beans, *J. Food Sci. Technology*, Vol. 20, pp. 64–67.

Ukhun, M.E., N.P. Okolie and A.D. Oyerinde (2005). Some mineral profiles of fresh and bottled palm wine–a comparative study, *Afr. J. Biotechnology*, Vol. 4 (8), pp. 829–832.

Ulmann, E. and C. Blaquiere (1972). Lindane: Monograph of an insecticide. *Freiberg im Breisgau: Schillinger*, pp. 6–65.

Ulrici, W. (2000). Contaminant soil areas, different countries and contaminant monitoring of contaminants in Environmental Process II. Soil Decontamination Biotechnology, *In*: H.J. Rehmand, and G. Reed (*ed.*), Wiley–VCH, Weinheim, FRG, pp. 5–42.

UNEP (2009). Report of the conference of the parties of the stockholm convention on persistent organic pollutants on the work of its fourth meeting. Convention on persistent–organic pollutants IV meeting, Geneva. *United Nations Environment Programme* http://chm.pops.int/Portals/0/Repository/COP4/UNEP–POPS–COP.4–38.English.pdf.

UNEP Chemicals (2005). Ridding the world of POPs: A guide to the Stockholm Convention on Persistent Organic Pollutants, *The Secretariat of the Stockholm Convention and UNEP's Information Unit for Conventions.* http://www.pops.int/documents/guidance/beg_guide.pdf.

USEPA (1983). Guidance for the reregistration of pesticide products containing chlorobenzilate as the active ingredient, *Environmental Protection Agency.* Washington, DC, pp. 6–7.

USEPA (2012). Integrated Pest Management (IPM) Principles, *Pesticides: Topical and Chemical Fact Sheets,* http://www.epa.gov/agriculture/tipm.html.

USFDA (2009). FDA Public Health Advisory: Safety of topical lindane products for the treatment of scabies and lice. http://www.fda.gov/Drugs/DrugSafety/PostmarketDrugSafetyInformationforPatientsandProviders/ucm110845.htm.

Van Eerd, L.L., R.E. Hoagland and J.C. Hall (2003). Pesticide metabolism in plants and microorganisms, *Weed Science,* Vol. 51, pp. 472–495.

Van Hamme, J.D., A. Singh and O.P. Ward (2003). Physiological aspects.part 1 in series of papers devoted to biosurfactants in microbiology and biotechnology, *Biotechnol. Advances,* Vol. 24, pp. 604–620.

Vandevivere P.C., R. Bianchi and W. Verstraete (1998). Treatment and reuse of wastewater from the textile wet–processing industry: Review of emerging technologies, *J. Chem. Technol. Biotechnology,* Vol. 72, pp. 289–302.

Vankar, P.S. and S. Ramashanker (2011). Lindane levels in the dumping grounds in Chinhat area, Lucknow India, *Elec. J. Environ. Agri. Food. Chemistry,* Vol. 10, No. 4, pp. 2081–2089.

Vasiliadou I.A., R. Molina, F. Martínez and J.A.Melero (2013) Biological removal of pharmaceutical and personal care products by a mixed microbial culture: Sorption, desorption and biodegradation, *Biochem. Eng. J.* Vol. 81, pp. 108–119.

Vasudevan, N. and P. Rajaram (2001). Bioremediation of oil sludge–contaminated soil, *Environ. International,* Vol. 26, pp. 409–411.

Vasudevan, P., V. Padmavathy and S.C. Dhingra (2003). Kinetics of biosorption of cadmium on baker's yeast, *Bioresour. Technology,* Vol. 89, pp. 281–287.

Veglio, F. and F. Beolchini (1997). Removal of metals by biosorption: A review, *Hydrometallurgy,* Vol. 44, pp. 301–316.

Vidali, M. (2001). Bioremediation. An overview, Pure Appl. Chemistry, Vol. 73, No. 7, pp. 1163–1172.

Vieno, N.M., H. Hrkki, T. Tuhkanen, T and Kronberg, L.(2007) Occurrence of pharmaceuticals in river water and their elimination in a pilot–scale drinking water treatment plant. *Environ. Sci. Technology,* Vol. 41, pp. 5077–5084.

Vigneswaran, S., H.N. Huu, D.S. Chaudhary and Y.T. Hung (2005). Physico-chemical Treatment Processes, Handbook of Environmental Engineering, Vol. 3, pp. 635–676.

Vijayaraghavan, K. and Y.S. Yun (2007). Chemical modification and immobilization of *Corynebacterium glutamicum* for biosorption of Reactive black 5 from aqueous solution, *Ind. Eng. Chem. Research*, Vol. 46, pp. 608–617.

Vitor, V. and C.R. Corso (2008). Decolorization of textile dye by *Candida albicans* isolated from industrial effluents, *J. Ind. Microbiol. Biotechnology*, Vol. 35, pp. 1353–1357.

Volesky, B. (Ed.), (1990). Biosorption and Biosorbents, In: Biosorption of Heavy Metals, CRC Press, Boca Raton, Florida, pp. 3–5.

Volesky, B. (2007). Biosorption and me, *Water Research*, Vol. 41, pp. 4017–4029.

Volesky, B. and I. Prasetyo (1994). Cadmium removal in a biosorption column, *Biotechnol. Bioengineering*, Vol. 43, pp. 1010–1015.

Von Fahnestock, F.M., G. B. Wickramanayake, K. J. Kratzke and W. R. Major (1998). *Biopile Design, Operation, and Maintenance Handbook for Treating Hydrocarbon Contaminated Soil*, Battelle Press, Columbus.

Wagner, R.D., S.J. Johnson, C.E. Cerniglia and B.D. Erickson (2011). Bovine intestinal bacteria inactivate and degrade deftiofur and ceftriaxone with multiple β–lactamases. *Antimicrob Agents Theraphy*, Vol. 11, pp. 4990–4998.

Wang, B.E. and Y.Y. Hu (2008). Bioaccumulation versus adsorption of reactive dye by immobilized growing *Aspergillus fumigatus* beads, *J. Hazard. Materials*, Vol. 157, pp. 1–7.

Wang, H. and K.Helliwell (2000). Epimerisation of catechins in green tea infusions, *Food Chemistry*, Vol. 70, pp. 337–344.

Wang, J.L. and C. Chen (2006). Biosorption of heavy metals by *Saccharomyces cerevisiae*: A review, *Biotechnol Advances*, Vol. 24, pp. 427–451.

Wang, L.K., D.A. Vaccari, Y. Li and N.K. Shammas (2004). Chemical precipitation. In: Wang, L.K., Hung, Y.T., Shammas, N.K. (Eds.), *Physicochemical Treatment Processes*, p. 141, Humana Press, New Jersey.

Watkinson, A.J., E.J. Murby, D.W. Kolpin and S.D. Constanzo (2009). The occurrence of antibiotics in an urban watershed: From wastewater to drinking water, *Sci. Total Environment*, Vol. 407, pp. 2711–2723.

Weigel, S., U. Berger, E. Jensen, R. Kallenborn, H. Thoresen, H. Hu¨hnerfuss (2004). Determination of selected pharmaceuticals and caffeine in sewage and seawater from Tromsø/Norway with emphasis on ibuprofen and its metabolites, *Chemosphere*, Vol. 56, pp. 583–592.

WHO (2003). Health risks of persistent organic pollutants from long–range trans–boundary air pollution, Chapter 3. Hexachlorocyclohexanes, Joint WHO/Convention Task Force on the Health Aspects of Air Pollution; *WHO: Geneva, Switzerland*, pp. 61–85.

Wijetunga, S., L. Xiufen, R. Wenquan and J. Chen (2007). Removal mechanisms of acid dyes of different chemical groups under anaerobic mixed culture, *Ruhuna J. Science*, Vol. 2, pp. 96–110.

Wilbourn, R.G., J.A. Newbour and J.T. Schofield (1994). Treatment of hazardous wastes using the thermatrix treatment system. *Pro. Therm. Treat. Radioact., Haz. Chem., Mixed, Munitions, Pharm. Wastes*. University of California, Irvine, pp. 221–223.

Wingenfelder, U., C. Hansen, G. Furrer and R. Schulin (2005). Removal of heavy metals from mine water by natural zeolites, *Environ. Sci. Technol*ogy, Vol. 39, pp. 4606–4613.

Wong K.K, C.K. Lee, K.S.Low and M.J. Haron (2003). Removal of Cu and Pb by tartaric acid modified rice husk from aqueous solutions, *Chemosphere*, Vol. 50, pp. 23–28.

Wu, J. and H.Q. Yu (2007). Biosorption of 2, 4–dichlorophenol by immobilized white–rot fungus *Phanerochaete chrysosporium* from aqueous solutions, *Bioresour. Technology*, Vol. 98, ppp. 253–259.

Wyszkowska, J., J. Kucharski and E. Waldowska (2002). The influence of diesel oil contamination on soil enzymes activity, *Rostlinna. Vyroba*, Vol. 48, pp. 58–62.

Xu, W.H., G. Zhang, S.C. Zou, X.D. Li and Y.C. Liu (2007). Determination of selected antibiotics in the Victoria Harbour and the Pearl River, South China using high performance liquid chromatography electrospray ionization tandem mass spectrometry, *Environ. Pollution*, Vol. 145, pp. 672–679.

Yadav, A., K. Chaudhary and P. Tauro (1996). Selection of dual purpose *Saccharomyces cerevisiae* strains for bakeries and distilleries, *Ind. J. Microbiology*, Vol. 36, pp. 119–120.

Yadav, S.K. and P.S. Ahuja (2007). Towards generating caffeine–free tea by metabolic engineering, *Plant Foods Hum. Nutrition*, Vol. 62, pp. 185–191.

Yalova, M. D., Dere, K.ve Deniz and S. K. Tanisi (2003). Caffeine in the stream, well and sea water of Yalova, Marmara Sea, Turkey, *Turkish J. Marine Sciences*, Vol. 9, pp. 179–186.

Yang, Q., A. Yediler, M. Yang and A. Kettrup (2005). Decolorization of an azo dye, reactive black 5 and mnP production by yeast isolate: *Debaryomyces polymorphus*, *Biochem. Engineering*, Vol. 24, pp. 249–253.

Yang, R., G. Ji, Q. Zhoe, C. Yaun and J. Shi (2005). Occurrence and distribution of organochlorine pesticides (HCH and DDT) in sediments collected from East China Sea, *Environ Int*ernational Vol. 31, pp. 799–804.

Yasuda, M., A. Miwa and Kitagawa (1995). Morphometric studies of renal lesions in 'itai itai' disease; chronic cadmium nephropathy, *Nephron*, Vol. 69, pp. 14–19.

Ye, J.H, J. Jin, H.L. Liang, J.L. Lu, YY. Du, X.Q. Zheng and Y.R. Liang (2009). Using tea stalk lignocellulose as an adsorbent for separating decaffeinated tea catechins, *Bioresour. Technology*, Vol. 100, pp. 622–628.

Yiruhan, Wang, Q.–J., Mo, C.–H., Li, Y.–W., Gao, P., Tai, Y.–P., Zhang, Y., Ruan, Z., Xu, J.–W., (2010). Determination of four fluoroquinolone antibiotics in tap water in Guangzhou and Macao, *Environ. Pollution*, Vol. 158, pp. 2350–2358.

Young, R.V. (ed.), (2000). World of Chemistry, Michigan, Gale Group.

Yu, Z. and X. Wen (2005). Screening and identification of yeasts for decolorizing synthetic dyes in industrial wastewater, *Int. Biodeterior. Biodegradation*, Vol. 56, pp. 109–114.

Zhang, A., L. Cui, G. Pan, L. Li, Q. Hussain, X. Zhang, J. Zheng and D. Crowley (2010). Effect of biochar amendment on yield and methane and nitrous oxide emissions from a rice paddy from Tai Lake plain, China, *Agr. Ecosyst Environment*, Vol. 139, pp. 469–475.

Zhang, H., H. Wan, L. Song, H. Jiang, H. Wang and C. Qiao (2010a). Development of an autofluorescent *Pseudomonas nitroreducens* with dehydrochlorinase activity for efficient mineralization of γ–hexachlorocyclohexane (γ–HCH), *J. Biotechnology*, Vol. 146, No. 3, pp. 114–119.

Zhemin, S., W. Wenhua, J. Jinping, Y. Jianchang, F. Xue and P. An (2001). Degradation of dye solution by an activated carbon fiber electrode electrolysis system, *J. Hazard. Materials*, Vol. 84, pp. 107–116.

Zheng, G., A. Selvam and J.W. Wong (2012a). Oil–in–water microemulsions enhance the biodegradation of DDT by *Phanerochaete chrysosporium*, *Bioresour. Technology*, Vol. 126, pp. 397–403.

Zhou, W. and W. Zimmerman (1993). Decolorization of industrial effluents containing reactive dyes by actinomycetes, *FEMS Microbiol. Letters*, Vol. 107, pp. 157–162.

Zhu, Y.M., D.Q. Zhou and D.Z. Wei (2004). Biosorption of Hg^{2+} by *Saccharomyces cerevisiae*, *J. Northeastern University*, Vol. 25, pp. 89–91.

Subject Index

B

D

E

F

Functional group 28, 29, 45, 46, 47, 49, 52, 67, 117, 148

G

Q

R

S